MODELING THE EFFECT OF DAMAGE IN COMPOSITE STRUCTURES

Aerospace Series List

MODELING THE EFFECT OF DAMAGE IN COMPOSITE STRUCTURES

SIMPLIFIED APPROACHES

Christos Kassapoglou

Delft University of Technology, The Netherlands

This edition first published 2015
© 2015 John Wiley & Sons Ltd

Registered office
John Wiley & Sons Ltd, The Atrium, Southern Gate, Chichester, West Sussex, PO19 8SQ, United Kingdom

For details of our global editorial offices, for customer services and for information about how to apply for permission to reuse the copyright material in this book please see our website at www.wiley.com.

Library of Congress Cataloging-in-Publication Data

applied for.

A catalogue record for this book is available from the British Library.

ISBN: 9781119013211

Typeset in 11/13pt TimesLTStd by Laserwords Private Limited, Chennai, India
Printed and bound in Singapore by Markono Print Media Pte Ltd

1   2015

Contents

Series Preface

The field of aerospace is multi-disciplinary and wide ranging, covering a large variety of products, disciplines and domains, not merely in engineering but in many related supporting activities. These combine to enable the aerospace industry to produce exciting and technologically advanced vehicles. The wealth of knowledge and experience that has been gained by expert practitioners in the various aerospace fields needs to be passed onto others working in the industry, including those just entering from University.

The *Aerospace Series* aims to be a practical, topical and relevant series of books aimed at people working in the aerospace industry, including engineering professionals and operators, allied professions such commercial and legal executives, and also engineers in academia. The range of topics is intended to be wide ranging, covering design and development, manufacture, operation and support of aircraft, as well as topics such as infrastructure operations and developments in research and technology.

Composite materials are being used increasingly in aerospace structures due to their impressive strength to weight properties, and the ability to manufacture complex integral components. However, in structural design, it is also important to be able to account for the effect of damage on the structure's integrity throughout its lifetime.

This book, *Modelling the Effect of Damage in Composite Structures: Simplified Approaches*, considers the various types of damage that can occur in composite structures and how they can be modelled during preliminary structural design. Analytical models are developed in order to understand and predict the physical phenomena that lead to the onset of damage and its evolution. The techniques provide a set of design tools and rules of thumb for a range of different types of damage in composite structures. The book complements the author's previous book *Design and Analysis of Composite Structures: With Application to Aerospace Structures, Second Edition* which is also in the Wiley Aerospace Series.

Peter Belobaba, Jonathan Cooper and Allan Seabridge

Preface

One of the main features of a good structural design is the ability to account for damage representative of what may occur during the lifetime of the structure and ensure that performance is not compromised with such damage present. For the case of airframe structures, knowledge of the effects of damage on structural performance is crucial in designing safe but weight-efficient structures.

This book provides a brief discussion of various types of damage and how they can be modelled during preliminary design and analysis of composite structures. It is addressed to graduate-level students and entry-level design and structural engineers. It is the result of a graduate course at Delft University of Technology covering design and analysis of composite structures in the presence of damage. While the emphasis is on aerospace structures, the principles and methods are applicable to all other fields with appropriate adjustment of safety factors and design criteria (e.g. impact damage).

Because the emphasis is on preliminary design and analysis, more accurate, and absolutely necessary during detailed design, computational methods such as finite elements are only briefly touched upon during the discussion. As a result, the accuracy and applicability of some of the approaches presented are not as good as those resulting from detailed finite element analysis. However, the methods are very efficient and provide good starting designs for more detailed subsequent evaluation. They can also be used to compare different designs and thus can be very useful for optimisation.

In a sense, this book is a natural continuation of my previous book on Design and Analysis of Composite Structures. There, the effect of damage was conservatively accounted for by applying appropriate knockdown factors on the allowable strength. Here, an attempt is made to replace these knockdown factors with analytical models that focus on understanding better some of the physical phenomena behind damage creation and evolution, while, at the same time, removing some of the conservatism associated with knockdown factors.

It is also recognised that some of the subjects discussed in this book such as the analysis of structures with impact damage or the fatigue analysis of composite structures are very much the subject of on-going research. As such, the methods presented here should be viewed as good design tools that may be superseded in the future as our understanding of damage creation and evolution in composite structures improves.

Chapter 1 includes a brief overview of types of damage and points out some important characteristics specific to composites manifested by increased notch sensitivity compared to metals. Chapter 2 discusses the effect of holes and provides improved methodology to obtain reliable failure predictions. Chapter 3 discusses through-thickness cracks and gives simple methods of analysis to obtain failure predictions. Delaminations are discussed in Chapter 4 where solutions for different structural details are given. Impact damage, which includes all previous types of damage, matrix cracks, holes (for high impact energies) and delaminations, is addressed in Chapter 5. A brief discussion of fatigue of composite materials with an emphasis on analytical models for predicting cycles to failure is given in Chapter 6. Constant amplitude and spectrum loading are discussed with an emphasis on how damage at different length scales may be accounted for in the analysis. Finally, design guidelines and rules of thumb that can be deduced from all previous chapters are summarised in Chapter 7.

Christos Kassapoglou

1

Damage in Composite Structures: Notch Sensitivity

1.1 Introduction

Owing to its construction, where two basic constituents, fibres and matrix, are combined, a composite structure shows a wide variety of types of damage. Damage may be specific to one or both of the constituents or involve interaction of the two. Furthermore, depending on the scale over which phenomena are described, damage may have different forms ranging from micro-voids or inconsistencies and cracks of the fibre/matrix interphase to large-scale delaminations, holes and laminate failures.

Here, the emphasis is placed on damage that is no smaller than a few fibre diameters with the understanding that this damage most likely is the result of creation and coalescence of damage at smaller scales, which are beyond the scope of this book. Within this framework, the most common forms of damage are matrix cracks, fibre/matrix interface failures, fibre failures, through-thickness failures (holes and cracks) and inter-ply failures such as delaminations. Of course any combination of these may also occur as in cases of impact damage. Representative forms of damage and their corresponding scales are shown in Figure 1.1.

In advanced composites typical of aerospace structures, the matrix has much lower strength than the fibres. Failure then typically initiates in the matrix and the associated damage is in the form of matrix cracks. These cracks usually appear in plies with fibres not aligned with the directions along which appreciable loads are applied [1]. Matrix cracks may also be present in a composite right after curing due to curing stresses [2] or tooling problems where heat uptake or cool-down during the cure cycle is not uniform [3].

This does not mean that damage may not initiate at a location where a small flaw (resin-rich region, resin-poor region, void and contamination) is present. Ideally, a damage model should start at the lowest possible scale where damage initiated and track the latter as it evolves and grows. As can be seen from Figure 1.1, however, this process may require bridging at least three to four orders of magnitude in the length scale. This means that separate models for the individual constituents are needed at the

Modeling the Effect of Damage in Composite Structures: Simplified Approaches, First Edition. Christos Kassapoglou.
© 2015 John Wiley & Sons, Ltd. Published 2015 by John Wiley & Sons, Ltd.

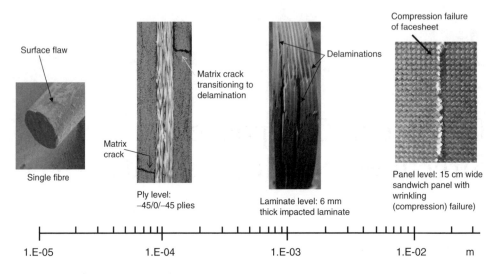

Figure 1.1   Typical damage at various scales of a composite structure

lower scales at which even the material homogeneity is in doubt. To minimise computational complexity, models that address macroscopic structures start at larger scales, the ply level or, less frequently, at somewhat lower scales and focus on aggregate flaws such as notches.

In general, a notch can be considered any type of local discontinuity such as a crack, hole, and indentation. Here, the definition of a notch is generalised and is not confined to a surface flaw. It can also be a through-the-thickness discontinuity. Notches act as stress risers and, as such, reduce the strength of a structure. The extent of the reduction is a function of the material and its ability to redistribute load around the notch. The possible range of behaviour is bounded by two extremes: (i) notch insensitivity and (ii) complete notch sensitivity.

1.2   Notch Insensitivity

This is the limiting behaviour of metals. Consider the notched plate at the top left of Figure 1.2. The shape and type of the notch are not important for the present discussion. Now assume that a purely elastic solution is obtained in the vicinity of the notch for a given far-field loading. Typically, there is a stress concentration factor k_t and, for an applied far-field stress σ, the stress at the edge of the notch is $k_t\sigma$. This is shown in the middle of Figure 1.2. If the material of the plate is metal, then, for sufficiently high values of the far-field stress σ, $k_t\sigma$ exceeds the yield stress σ_y of the material. As a first-order approximation, one can truncate the linear stress solution in the region where the local stress exceeds the yield stress (shown by a dashed line in the middle of Figure 1.2) by setting the stress there equal to the yield stress. To maintain force

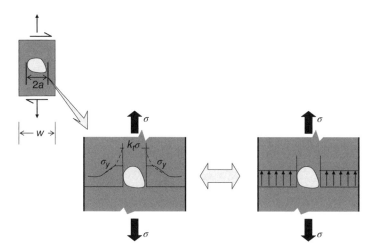

Figure 1.2 Stress distribution in the vicinity of a notch-insensitive material

equilibrium, the region where the stress equals σ_y must extend beyond the point of intersection of the horizontal line at σ_y and the linear stress solution such that the areas under the original curve corresponding to the linear solution and the modified 'truncated' curve are equal.

For sufficiently high σ and/or sufficiently low σ_y value, the material on either side of the notch yields and the stress distribution become the one shown on the right of Figure 1.2.

This means that the stress aligned with the load on either side of the notch is constant and there is no stress concentration effect any more. The stress is completely redistributed and only the reduced area due to the presence of the notch plays a role. More specifically, if F_{tu} is the failure strength of the material (units of stress), the force F_{fail} at which the plate fails is given by the material strength multiplied by the available cross-sectional area:

$$F_{fail} = F_{tu}(w - 2a)t \tag{1.1}$$

with w and $2a$ the plate and notch widths, respectively, and t the plate thickness.

At the far-field, the same force is given by

$$F_{fail} = \sigma w t \tag{1.2}$$

The right-hand sides of Equations 1.1 and 1.2 can be set equal and a solution for the far-field stress that causes failure can be obtained:

$$\sigma = F_{tu}\left(1 - \frac{2a}{w}\right) \tag{1.3}$$

A plot of the far-field stress as a function of normalised notch size $2a/w$ is shown in Figure 1.3. The straight line connecting the failure strength F_{tu} on the y-axis with the

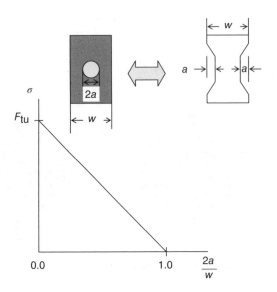

Figure 1.3   Notch-insensitive behaviour

point $2a = w$ on the *x*-axis gives the upper limit of material behaviour in the presence of a notch.

It should be pointed out that, for this limiting behaviour, the shape of the notch is not important. The specimen with the hole and the dog-bone specimen shown in Figure 1.3 are completely equivalent.

1.3   'Complete' Notch Sensitivity

At the other extreme of material behaviour are brittle materials that are notch-sensitive, such as some composites and ceramics. In this case, if there is a stress riser due to the presence of a notch with a stress concentration factor k_t, failure occurs as soon as the maximum stress in the structure reaches the ultimate strength of the material. For a far-field applied stress σ, this leads to the condition:

$$k_t \sigma = F_{tu} \tag{1.4}$$

The situation is shown in Figure 1.4. Here, there is no redistribution of stress in the vicinity of the notch. For the case of an infinite plate in Figure 1.4, or a very small notch, the far-field stress to cause failure is given by rearranging Equation 1.4:

$$\sigma = \frac{F_{tu}}{k_t} \tag{1.5}$$

For finite plates, with larger notches, finite width effects reduce further the strength of the plate. In the limit, as the notch size approaches the width of the plate, the strength goes to zero:

$$\sigma \to 0 \quad \text{as} \quad 2a \to w \tag{1.6}$$

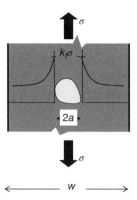

Figure 1.4 Stress distribution in the vicinity of a notch-sensitive material

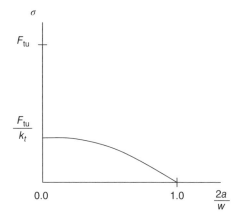

Figure 1.5 Notch-sensitive material

Equations 1.5 and 1.6 are combined in Figure 1.5, which shows the notch-sensitive behaviour.

1.4 Notch Sensitivity of Composite Materials

The types of behaviour discussed in the previous two sections are the two extremes that bracket all materials. It is interesting to see where typical composite materials lie with respect to these two extremes. Experimental data for various composite laminates with different hole sizes under tension are shown in Figure 1.6. The test data are taken from Ref. [4].

Note that two curves, very close to each other, are shown for the 'completely notch-sensitive' behaviour. One corresponds to the [15/−15]s and the other to the [15/−15/0]s laminate.

It is seen from Figure 1.6 that the composite data fall between the two curves. More importantly, even at very small holes, there is a significant drop of the strength towards

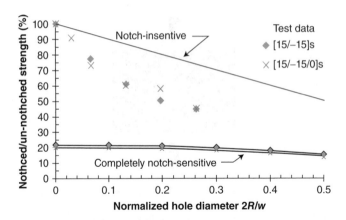

Figure 1.6   Test results for composite laminates with holes under tension

the curve of complete notch sensitivity and, at higher hole diameters $(2R/w > 0.7)$, the data tend to follow that curve. However, the fact that the data start at the top curve and drop towards the lower curve suggests that composites have some load redistribution around a notch but the redistribution is limited. A damage zone or process zone is created at the edge of the hole with matrix cracks, broken fibres and delaminations. This process zone limits the stress to a value equal or close to the undamaged failure strength. As the load is increased, the stress inside the process zone stays constant. The size of the process zone increases and the strains in the material next to the hole increase. As the load is increased further, a point is reached where the structure can no longer store energy and fails. In general, therefore, composites are notch sensitive but they do have some limited ability to redistribute load around notches. This will be of some significance in subsequent chapters when stresses in the vicinity of a notch are discussed in more detail.

Exercises

1.1   Discuss how small-scale defects and flaws affect the scatter in the static strength of a composite structure. Then, discuss how the presence of damage of a size sufficient to drive failure may reduce the scatter of the static strength.

1.2   Often, but not always, the presence of a notch of sufficient size and severity, not only drives failure but also masks the effects of smaller more benign notches in a way that the scatter of test results is lower with a notch than without it. Table E1.1 gives the un-notched and notched strength values for [45/−45/0]s, [0/45/−45]s and [45/0/−45]s laminates. Lumping all laminates in one data set, determine the B- and A-basis for un-notched specimens and for each hole diameter. Do this as a fraction of the corresponding mean value. Comment on the knockdown due to the hole and how it relates to the scatter of the test data.

Table E1.1 Strength as a function of hole size

Tension strength (MPa)	Hole diameter (mm)			
	3.282	6.578	10.31	50.088
754	388	346	379	316
683	466	358	354	342
793	443	423	332	224
476	422	370	355	291
696	431	306	331	347
811	452	380	331	305
801	463	364	343	278
780	415	321	310	286
796	435	398	307	292
747	444	355	312	306
556	434	368	359	308
779	429	362	310	295
741	424	348	295	272
768	458	386	316	282
604	427	368	334	282

1.3 The following two curves for complete notch sensitivity of specimens with holes are given:

$$\frac{\sigma_0}{F_{tu}} = \frac{3(1 - (2a/w))}{[2 + (1 - (2a/w))^3]k_t^\infty}$$

and

$$\frac{\sigma_0}{F_{tu}} = \frac{2 - (2R/w)^2 - (2R/w)^4 + (2R/w)^6(k_t^\infty - 3)(1 - (2R/w)^2)}{2k_t^\infty}$$

Plot them in the same figure for a low k_t^∞ value of 2 and a high k_t^∞ value of 8. Discuss the differences.

References

[1] Highsmith, A.L. and Reifsnider, K.L. (1982) Stiffness-reduction mechanisms in composite laminates, in *Damage in Composite Materials* (ed K.L. Reifsnider), American Society for Testing and Materials, Philadelphia, PA, pp. 103–117, ASTM STP 775.

[2] Rohwer, K. and Jiu, X.M. (1986) Micromechanical curing stresses in CFRP. *Compos. Sci. Technol.*, **25**, 169–186.

[3] Penn, L.S., Chou, R.C.T., Wang, A.S.D. and Binienda, W.K. (1989) The effect of matrix shrinkage on damage accumulation in composites. *J. Compos. Mater.*, **23**, 570–586.

[4] Lagacé, P.A. (1982) Static tensile fracture of graphite/epoxy. PhD thesis. Massachusetts Institute of Technology.

2

Holes

One of the most common stress risers in a composite structure is a hole. The presence of holes is inevitable in an airframe. Even though assembly methods other than fastening are used through increased co-curing and adhesive bonding, some fastener holes are always present. In addition, lightening holes (to reduce weight), holes to accommodate systems equipment (hydraulics, electrical, etc.) and window and access door cutouts are ever present (Figure 2.1). Finally, holes (punctures) created by tool drops, runway debris or other forms of damage also occur during service.

Knowing how a composite structure responds in the presence of holes is, therefore, very important for generating robust and damage-tolerant designs. There is one additional reason for understanding how the strength of a composite structure is affected by the presence of a hole: Holes can be used to model other more complex forms of damage. Treating impact damage or through-thickness cracks as holes of equivalent size can simplify the design and analysis of such structures.

There are different types of holes depending on their shape and whether they are loaded or not. In addition to circular holes which will be the focus of this chapter, there are elliptical holes or cutouts and holes of irregular shape caused by high-speed damage (for example, ballistic damage). In terms of loading, one can distinguish between unloaded holes, fastener holes with bearing and torque-up loads and filled holes where the filler material is typically the shank of an un-torqued fastener which, being much stiffer, enforces a nearly circular shape on the hole circumference and does not allow it to elongate. These are all shown in Figure 2.2.

It is important to note that fastener holes tend to lose their torque-up over time. One extreme case is shown in Figure 2.3. Here, a nut and bolt were used to fasten two pieces of a (bismaleimide) composite plate. Prior to torquing-up the bolt, two back-to-back flat surfaces were machined on the bolt shank, and strain gauges were installed. The readings of the back-to-back strain gauges were monitored over time, and they are shown as a percentage of the initial strain reading in Figure 2.3.

Modeling the Effect of Damage in Composite Structures: Simplified Approaches, First Edition. Christos Kassapoglou.
© 2015 John Wiley & Sons, Ltd. Published 2015 by John Wiley & Sons, Ltd.

Figure 2.1   Fuselage structure with lightening and attachment holes

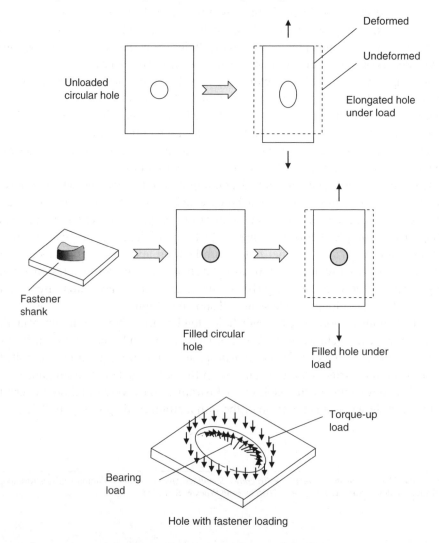

Figure 2.2   Loaded and unloaded holes

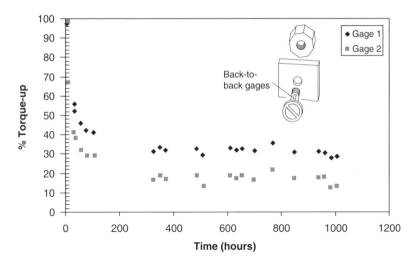

Figure 2.3   Loss of torque-up over time

The data in Figure 2.3 show a drastic reduction in the strain in the bolt as a function of time. This corresponds to a proportional loss of torque-up. Within 10 days, the torque is down to ~30% of its initial value. After that the reduction is much slower. These data correspond to a regular screw and nut. Typical fasteners for aircraft applications show significantly less reduction. Nevertheless, over the decades corresponding to the life of an aircraft, such loss of torque-up is not uncommon. Reduced torque-up translates to reduced bearing strength capability. For this reason, it is common to design bolted joints assuming very little or no torque values corresponding to 'finger-tight' fasteners.

In general, but not always, filled hole tension (FHT) is lower than open hole tension (OHT) [1] while filled hole compression (FHC) is higher than open hole compression (OHC). Designing conservatively in the presence of holes would then, at a minimum, require FHT and OHC tests to establish the lowest values to be expected in service. At the same time, FHT and FHC can be used to define the limits of a bearing-by-pass curve for a specific laminate, as shown schematically in Figure 2.4.

If there is a single member with a finger-tight fastener, that is, there is no second member that the fastener attaches to, then all the applied load P by-passes the fastener and comes out as load P at the other end of the same member. If the load is tensile, this corresponds to the FHT case. If it is compressive, it corresponds to the FHC case. Both are shown on the vertical axis of Figure 2.4.

In a single lap joint with a single fastener, all the load P in one bolted member is transmitted through the fastener to the other member. This means that there is no load by-passing the fastener. The entire load becomes bearing load or bearing stress, which corresponds to a point on the horizontal axis in Figure 2.4. The bearing stress $P/(Dt)$,

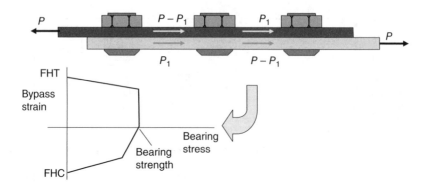

Figure 2.4   Notional bearing by-pass curve for a laminate

where D is the fastener diameter and t the thickness of the member in question, is to be compared to the bearing strength of the laminate used.

   If there are at least two fasteners connecting two members, then only a portion P_1 of the total load is transmitted to the other plate through the first fastener, as shown in Figure 2.4. The remaining load $P - P_1$ by-passes the first fastener and reaches the next fastener. If there are only two fasteners, all of it is transmitted through the second fastener to the second member. If there are more than two fasteners, then, again, a portion is transmitted through the second fastener and the remainder by-passes to the next fastener. For each of these 'interior' fasteners, there is a combination of bearing stress and by-pass strain that causes failure of the joint. For example, in Figure 2.4, the second fastener has a bearing stress $(P - 2P_1)/(Dt)$ since the net load transmitted through that fastener is $P - P_1$ (by-pass load after the first fastener) minus P_1 (by-pass load after the second fastener). The by-pass strain is P_1/E_m where E_m is the axial stiffness of the member.

   Analyses of holes with fastener loads have been presented by several authors and are not the focus of this chapter, for example [2–6] with design methods and guidelines [7–9]. The emphasis here will be on open holes. They correspond to a relatively simple case which can give insight into the more general case of the loaded hole. And, in some cases, open holes are convenient to use in damage tolerance programs of composite structures. First because, as already mentioned, other types of damage such as cracks or impact damage may conservatively be modelled as holes of equivalent sizes (see, for example, Section 5.6.1), and, second, because they can help establish limit load capability of a structure. A small hole, typically 6.35 mm in diameter, is considered visible damage and, as such, it is a limit load requirement for a composite structure. Certification of a composite structure requires demonstration that the structure can meet limit load in the presence of visible damage. Therefore, even if a hole is not present in the structure, showing that the structure can meet limit load with a 6.35-mm hole can help demonstrate its damage tolerance capability.

2.1 Stresses around Holes

The case of a homogeneous orthotropic laminate with an elliptical hole under tension has been treated by Lekhnitskii [10] and for a wide variety of cut-outs by Savin [11]. Stress functions and complex elasticity are used to obtain the in-plane stresses in an infinite plate. The laminate is assumed to be symmetric, so no bending–stretching coupling occurs.

Of interest for predicting failure is the magnitude of the circumferential stress σ_θ, which can be shown to be

$$\sigma_\theta(r=a) = \frac{-K\cos^2\theta + (1 + \sqrt{2K-m})\sin^2\theta}{\sin^4\theta - m\sin^2\theta\cos^2\theta + K^2\cos^4\theta}\sigma \tag{2.1}$$

with

$$K = \sqrt{\frac{E_{11}^L}{E_{22}^L}} \tag{2.2}$$

$$m = 2v_{12}^L - \frac{E_{11}^L}{G_{12}^L} \tag{2.3}$$

In the above, σ is the far-field applied stress, a is the radius of the hole, and E_{11}^L, E_{22}^L, G_{12}^L and v_{12}^L are elastic constants for the entire homogenised laminate. The situation is shown in Figure 2.5.

The quantity on the right-hand side of Equation 2.1 multiplying the far field stress σ is the stress concentration factor (SCF). The laminate's elastic constants are given by

$$E_{11}^L = \frac{1}{ha_{11}}$$

$$E_{22}^L = \frac{1}{ha_{22}}$$

$$G_{12}^L = \frac{1}{ha_{66}}$$

$$v_{12}^L = -\frac{a_{12}}{a_{11}} \tag{2.4a–d}$$

where h is the laminate thickness and a_{ij} are the corresponding entries of the inverse of the A matrix of the laminate.

Note that, because of symmetry, only the region $0 \leq \theta \leq 90$ is of interest. For the case of a quasi-isotropic lay-up

$$E_{11}^L = E_{22}^L \quad \text{and}$$

$$G_{12}^L = G = \frac{E}{2(1 + v_{12}^L)}$$

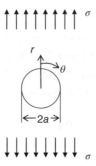

Figure 2.5   Circular hole in an infinite composite plate under tension

Substituting in Equations 2.2 and 2.3 gives

$$K = 1$$

$$m = 2v_{12}^L - \frac{E_{11}^L}{E_{11}^L / \left(2\left(1 + v_{12}^L\right)\right)} = -2$$

Then, substituting in the right-hand side of Equation 2.1 gives the well-known results for the SCF for a quasi-isotropic lay-up:

$$
\begin{aligned}
\text{SCF} &= \frac{-(1)\cos^2\theta + \left(1 + \sqrt{2 - (-2)}\right)\sin^2\theta}{\sin^4\theta - (-2)\sin^2\theta\cos^2\theta + (1)\cos^4\theta} \\
&= \frac{\cos^2\theta + 3\sin^2\theta}{\sin^4\theta + 4\sin^2\theta\cos^2\theta + \cos^4\theta}\text{(quasi-isotropic)}
\end{aligned}
$$

It is easy to see that the above quantity is maximised when $\theta = 90$, giving the well-known result SCF = 3 for quasi-isotropic and isotropic materials.

For an orthotropic plate, the θ value at which SCF is maximised depends on the lay-up. Results for some laminates for a typical thermoset material are shown in Figure 2.6.

The stacking sequences of the laminates in Figure 2.6 are given only in terms of percentages of 0, 45/−45 and 90° plies. This is done because the actual stacking through the thickness does not matter since Equations 2.1–2.4a–d depend only on the A matrix which is independent of the stacking sequence. This is a direct result of the assumption that the laminate is homogeneous.

It can be seen from Figure 2.6 that the maximum value of the SCF can be much higher than the typical value of 3 for metals and it is maximised for laminates consisting only of 0° plies. It occurs at $\theta = 90°$. As the number of 45° plies increases, the maximum SCF decreases. For 45/−45 dominated laminates, the SCF moves away from the $\theta = 90°$ location in the region between 50° and 65°. The maximum SCF value

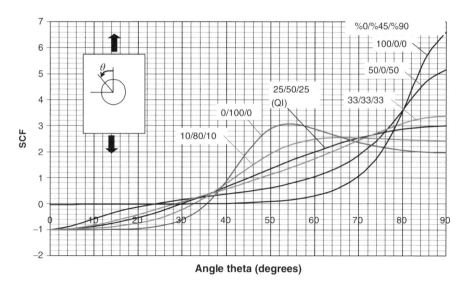

Figure 2.6   Stress concentration factor around a hole in an orthotropic plate

for some of these cases, see the 10/80/10 laminate in Figure 2.6, can be less than 3. As expected from the fact that the laminate is locally in compression there, the SCF is negative near $\theta = 0°$.

As already mentioned, the SCF determined here is independent of the stacking sequence. However, the presence of a free edge at the hole boundary will create inter-laminar stresses whose magnitude will strongly depend on the stacking sequence. In fact, for certain lay-ups, failure near at the hole edge may start as delaminations caused by these inter-laminar stresses. Methods of determining inter-laminar stresses near hole boundaries can be found in [12–14] based on finite elements. An excellent semi-analytical approach can be found in [15].

Equations 2.1–2.3 show that the SCF is independent of the hole size. This should have been expected for an infinite plate. For a finite plate, since the hole can be treated as a notch, the discussion in the previous chapter suggests that there is a strong dependence on the hole size. Detailed solutions that account for the effect of finite width can be found in the references, for example, [4, 9]. Accurate approximations of the finite width effect were obtained by Tan [16]. He provided various analytical expressions for the finite width correction factor for a circular hole. The simplest to use is

$$k_t = \frac{2 + (1 - (2a/w))^3}{3(1 - (2a/w))} k_t^\infty \tag{2.5}$$

where w is the plate width and $2a$ is the hole diameter, as shown in Figure 2.7. This factor has been found to be within 5–10% of more accurate analysis methods.

The effect of the ratio of the hole diameter to the plate width is shown in Figure 2.7, where the finite width correction factor k_t normalised by the SCF for an infinite plate

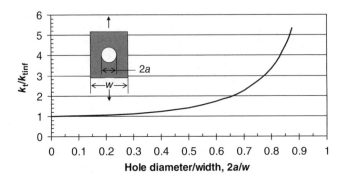

Figure 2.7   Circular hole in a finite width plate under tension

is plotted. For small diameter holes or large values of the width w, k_t is very nearly equal to 1. However, it rises quickly as $2a/w$ increases beyond 0.5.

For a plate of finite width, the SCF can be obtained by multiplying the SCF in the right-hand side of Equation 2.1 with the right-hand side of Equation 2.5.

If the solution for a plate with a hole under uniaxial load is known, solutions for combined load cases, such as biaxial loading or shear, can be obtained through superposition (see Exercise 2.8).

2.2   Using the Anisotropic Elasticity Solution to Predict Failure

The results so far are valid for both tension and compression loading. If composites were completely notch-sensitive, the failure stress for a plate with a hole would be given by dividing the undamaged strength, for tension or compression, respectively, by the SCF accounting for finite width effects. As was mentioned in the previous chapter (see also Figure 1.6), this is not the case. Some load redistribution around the hole occurs when damage starts and grows, allowing the laminate to withstand significantly higher loads than that. Comparisons of analytical predictions to test results for different lay-ups, unidirectional tape or fabric materials, loadings and environmental conditions are shown in Table 2.1. In that table, CTA, RTA and ETW stand for cold temperature ambient ($-54\,^\circ$C), room temperature ambient and elevated temperature wet ($82\,^\circ$C saturated with moisture), respectively.

The predicted SCF in Table 2.1 is obtained using Equations 2.1 and 2.5:

$$\text{SCF} = \frac{2 + (1 - (2a/w))^3}{3(1 - (2a/w))} \frac{-K\cos^2\theta + \left(1 + \sqrt{2K - m}\right)\sin^2\theta}{\sin^4\theta - m\sin^2\theta\cos^2\theta + K^4\cos^4\theta} \qquad (2.6)$$

The test coupons used in Table 2.1 had a $2a/w$ of $\sim$0.12, giving a finite width correction factor of 1.015.

The last column in Table 2.1 shows the percenatge difference between theoretical prediction and experimental results. The largest and smallest deviations from test

Table 2.1 Predicted and experimental SCF

Lay-up[a]	Environment	T or C	SCF theoretical	SCF experimental	% difference
50/0/50 tape	CTA	C	5.07	2.36	114.8
50/0/50 tape	RTA	C	5.07	2.66	90.6
50/0/50 tape	ETW	C	5.07	2.07	144.9
50/0/50 fabric	CTA	T	5.16	2.32	122.4
50/0/50 fabric	RTA	T	5.16	2.39	115.9
50/0/50 fabric	CTA	C	5.16	2.31	123.4
50/0/50 fabric	RTA	C	5.16	2.44	111.5
50/0/50 fabric	ETW	C	5.16	2.40	115.0
20/50/25 fabric	CTA	T	3.00	1.66	80.7
20/50/25 fabric	RTA	T	3.00	1.81	65.7
20/50/25 fabric	ETW	T	3.00	1.70	76.5
10/80/10 fabric	RTA	T	2.55	1.44	77.1
0/100/0 fabric	CTA	T	3.08	1.04	196.2
0/100/0 fabric	RTA	T	3.08	1.08	185.2
0/100/0 fabric	ETW	T	3.08	1.03	199.0

T, tensile; C, compressive.
[a] %0/%45/%90.

results are highlighted in that column. It is seen that the closest to theoretical results is 66% for quasi-isotropic (25/50/25) RTA tension (*T*). The largest discrepancy occurs for 100% 45 laminate where the prediction is three times higher than the test result.

The discrepancies between tests and analytical predictions are approximately equally large for tension and compression and are independent of environment. It should be noted that the predictions for different environments use the same environmental factors in E_{11}, E_{22} and G_{12}, and this is why they are identical across environments. The test results show a small dependence on the environment. Interestingly, ETW appears to have a lower experimentally measured SCF than CTA or RTA in most cases. Also, SCF for CTA is lower than SCF for RTA environment.

One final important conclusion that can be drawn from Table 2.1 is in order. The lowest experimentally measured SCF occurs for all ±45 laminates and is very close to 1. This means that all ±45 laminates are very effective in transferring the load around holes with almost no stress concentration. This will also be of some importance later in Chapter 5 when impact will be discussed.

It is obvious from the results of Table 2.1 that an approach to predict failure of a laminate with a hole based on the analytically determined SCF is very conservative. More elaborate methods accounting for the damage created and how it affects load redistribution are necessary for improved predictions. Some of these will be discussed below.

2.3 The Role of the Damage Zone Created Near a Hole

Consider a coupon with a hole under tension, as shown in Figure 2.8. For simplicity, assume that the highest SCF around the hole occurs at $\theta = 90°$.

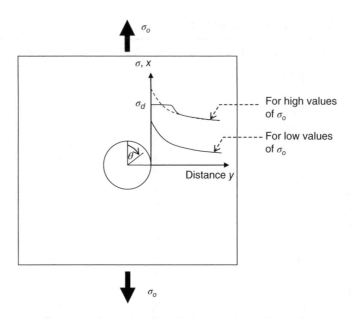

Figure 2.8   Stress distribution near the edge of a hole

For low applied far-field stress σ_o, the peak stress at the hole edge is lower than the undamaged failure strength σ_d, and the linear solution predicted by Lekhnitskii [10] is valid. This is the lower continuous curve in Figure 2.8. As the far-field applied stress increases, a point is reached when the peak stress at the hole edge equals the undamaged failure strength σ_d. Damage starts at that location and, as the load increases further, grows without necessarily causing final failure of the laminate. A damage or process zone is created which allows some load redistribution around the hole and prevents the stress from reaching the magnitude predicted by the SCF calculation. The stress in the damaged region remains constant and equal to σ_d. The stress distribution as a function of distance from the hole edge no longer follows the linear solution. This is shown as the upper continuous curve in Figure 2.8. The dashed curve denotes the linear solution for that case. It is expected that, outside the damaged region, the stress distribution will transition from σ_d to the linear stress distribution.

Depending on the laminate stacking sequence, the type of damage created is different. Two extreme examples are shown schematically in Figure 2.9.

A laminate with all unidirectional plies with the fibres aligned with the load is shown in Figure 2.9a. Upon loading, the matrix next to the hole edge cracks. Matrix cracks parallel to the fibres are created (fibre splits), which may extend all the way to the loaded edges of the specimen. At the other end of the spectrum, matrix-dominated laminates have very few or no fibres along the direction of the loading. Such a laminate, consisting only of 45, −45 and 90° plies is shown in Figure 2.9b. Upon loading, the matrix fails at the hole edge, but the presence of multiple fibre directions through the thickness leads to some fibre splits along fibres in the 45 and −45 plies

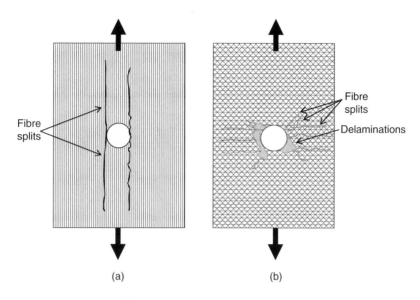

Figure 2.9   Failure modes in laminates with holes under tension

and some along the fibres in the 90° plies. These matrix cracks terminate in the through-the-thickness direction when a ply of different orientation is encountered. However, when a crack in one ply meets one in another ply, for example a 45° crack meets a 90° crack at a 45/90 ply interface, delaminations occur. Therefore, the laminate in Figure 2.9b consists of a combination of short matrix cracks in the 45 and −45 plies, longer cracks in the 90° plies and delaminations at various ply interfaces.

It can be seen from this brief qualitative discussion that the type of damage created near a hole can be very complex and is a strong function of the stacking sequence. This gets further complicated when significant inter-laminar stresses are present at the hole edge. In fact, damage creation and evolution spans many length scales from micro voids triggering a matrix crack in the nano scale to delaminations in the centimetre scale. Bridging these phenomena across scales is very challenging.

2.4   Simplified Approaches to Predict Failure in Laminates with Holes: the Whitney–Nuismer Criteria

The complexity of the damage makes it very difficult to model it accurately with relatively simple means. Early modelling approaches, such as the one by Gürdal and Haftka [17], showed that local failure phenomena such as fibre kinking play a major role in the creation of the process zone. More recently, promising methods that can capture different types of damage, their interaction and the local stress re-distribution have been developed [18], but they are computationally very intensive. For this reason, simplifications that acknowledge the presence of damage but are based on relatively simple approaches are sought. One of the best such approaches is the method proposed by Whitney and Nuismer [19].

The goal in their approach was to use the linear solution as the starting point but also acknowledge the presence of the damage region created next to the hole. In addition, they wanted to indirectly account for the load redistribution caused by the damage region.

It can be shown (see also Figure 2.8) that, for an infinite plate with a hole, the circumferential stress σ_θ at $\theta = 90°$ is given as a function of distance from the centre of the hole, to a very close approximation, by [16]

$$\sigma_x(x = 0, y) = \sigma_o \left[1 + \frac{1}{2}\left(\frac{R}{y}\right)^2 + \frac{3}{2}\left(\frac{R}{y}\right)^4 - (k_t^\infty - 3)\left(\frac{5}{2}\left(\frac{R}{y}\right)^6 - \frac{7}{2}\left(\frac{R}{y}\right)^8\right) \right]$$
(2.7)

where k_t^∞ is the SCF for a plate of infinite width.

On examining Equation 2.7, it can be seen that the only dependence on stacking sequence is through k_t^∞. This means that differences from one laminate to another at $\theta = 90°$ will be significant only when the last two terms of Equation 2.7 are significant. Given the fact that y is distance from the hole centre and thus is always greater than the hole radius R, the last two terms of Equation 2.7 are appreciable only very near the hole edge. At longer distances, R/y to the sixth or eighth power is negligible compared to the second or fourth power. This means that all laminates would collapse to a single curve far from the hole edge.

Note that this expression is valid only at $\theta = 90°$. If the highest stress occurs at a different location on the hole circumference, such as the all 45/−45 laminate in Figure 2.6, a different expression must be derived, or the complete complex elasticity solution must be used.

One experimental observation that needs to be accounted for is that larger holes lead to lower coupon strengths even for very wide specimens where the finite width effects are negligible. A hint that might help explain this behaviour is obtained from Equation 2.7 by plotting the stress distribution for two different hole radii and the same applied stress σ_o.

Arbitrarily, a hole with 5 mm radius and a hole with 15 mm radius are selected. The laminate is assumed to have $k_t^\infty = 4$. For each hole radius, the stress distribution as a function of distance from hole edge is plotted at $\theta = 90°$, having moved the centre of the smaller hole so that the hole edges coincide. The result is shown in Figure 2.10.

It can be seen from Figure 2.10 that, for both holes, the stress distributions start from the same value, the value of SCF = 4, and they go to the same value, SCF = 1, far from the hole. However, in between, there is an important difference. The stress distribution for the larger hole decays more slowly. This is significant because it means that the larger hole has a greater region of higher stresses next to it than the smaller hole. For example, a horizontal line at an edge stress equal to 2.5 times the far field stress in Figure 2.10 extends over a distance less than 1 mm from the hole edge for the small hole but over almost 2 mm from the hole edge for the larger hole. Therefore, the probability of some small flaw that can trigger final failure being found in a region of

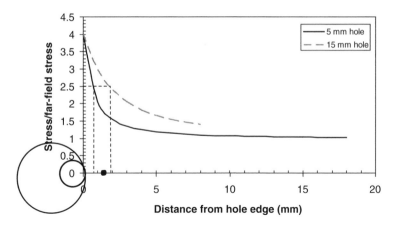

Figure 2.10 Stress distribution as a function of distance for 5 and 15 mm radius holes

high stresses is higher for the larger hole than for the smaller hole. As a result, higher holes have, on average, lower failure stresses.

An equivalent way to see this is by assuming that there is a flaw at a certain distance away from the hole edge, say at 1.5 mm. This is shown as a solid circle on the horizontal axis of Figure 2.10. This flaw is in a region of a much higher stress in the coupon with the larger hole, and thus will trigger failure at a lower far-field stress than in the coupon with the smaller hole.

It should be noted that the word 'flaw' was used here somewhat vaguely on purpose. It refers to any small-scale discontinuity (void) or inconsistency in the material (resin-rich or resin-poor region), or even small-scale contamination that may act as the trigger for final failure.

The above discussion shows the importance of accurate knowledge of the stresses in the hole's vicinity. Whitney and Nuismer proposed two ways to circumvent the fact that the damage created near the hole edge affects the stress distribution in a complex way. In both proposals they suggested abandoning the peak value at the hole edge as a reliable number because the damage created redistributes the stress. They proposed to use the linear solution to evaluate the stress at a 'characteristic distance' away from the hole edge, and, if that stress is greater than the undamaged failure strength of the laminate, the laminate fails. In their second approach, they proposed, instead of evaluating the stress at a point at some characteristic distance away, to average the stress over a characteristic distance (different from that used for the point stress evaluation). If that average stress exceeds the undamaged strength of the laminate, there is failure.

Denoting the characteristic distances by d_o for the point stress and a_o for the average stress criterion, the failure condition for each of the two cases can be written as (see also Figure 2.8 for coordinate system definition)

$$\sigma_x(x = 0, y = R + d_o) = \sigma_f \tag{2.8}$$

$$\frac{1}{a_o} \int_R^{R+a_o} \sigma_x(x = 0, y)\mathrm{d}y = \sigma_f \tag{2.9}$$

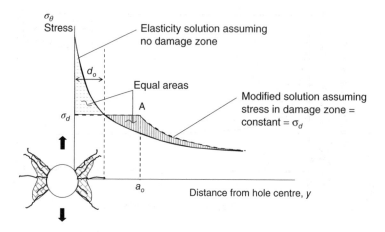

Figure 2.11   Modification of linear stress distribution near a hole edge

where σ_f is the undamaged failure strength and σ_x is given by Equation 2.7.

Both Equations 2.8 and 2.9 imply that the stress near the hole edge is not equal to the stress predicted by the linear solution. One interpretation of these equations is that they assume that failure occurs when the strain energy density stored in the damage region reaches a critical value. This can be seen by considering Figure 2.11.

Assume that the stress in the damage region is constant and equal to σ_d. It is expected that $\sigma_d \approx \sigma_f$. According to Equation 2.8, at failure, σ_θ at a distance d_o from the hole edge equals σ_f. This means that, at failure, the area under the curve obtained from the elasticity solution between R and $R + d_o$ is constant but for a multiplier σ_o:

$$\int_R^{R+d_o} \sigma_x(x=0,y)dy = \sigma_o \begin{bmatrix} d_o - \dfrac{R^2}{2}\left(\dfrac{1}{R+d_o} - \dfrac{1}{R}\right) - \dfrac{R^4}{2}\left(\dfrac{1}{(R+d_o)^3} - \dfrac{1}{R^3}\right) \\[2mm] - (k_t^\infty - 3)\left(-\dfrac{R^6}{2}\left(\dfrac{1}{(R+d_o)^5} - \dfrac{1}{R^5}\right)\right. \\[2mm] \left. + \dfrac{R^8}{2}\left(\dfrac{1}{(R+d_o)^7} - \dfrac{1}{R^7}\right)\right) \end{bmatrix}$$

$$(2.10)$$

Therefore, Equation 2.8 implies that the far-field stress to cause failure in a laminate with a hole is given by

$$\sigma_o = \frac{\sigma_f}{\begin{bmatrix} d_o - (R^2/2)\left((1/R + d_o) - (1/R)\right) - (R^4/2) \\ \cdot \left((1/(R+d_o)^3) - (1/R^3)\right) - (k_t^\infty - 3) \\ \cdot \left(-(R^6/2)((1/(R+d_o)^5) - (1/R^5)) + (R^8/2)((1/(R+d_o)^7) - (1/R^7))\right) \end{bmatrix}}$$

$$(2.11)$$

The implication of Equation 2.11 is that failure occurs when the total load per unit of local width reaches a critical value given by the right-hand side of Equation 2.11.

Equation 2.9 leads to a slightly different interpretation. The requirement that the stress in the damage region is constant and equal to σ_d truncates the stress versus distance curve of Figure 2.11. In addition, in order to maintain force equilibrium, the horizontal portion of the curve in Figure 2.11 must extend beyond the intersection of the horizontal and curved lines. To a first approximation, the new stress distribution can be approximated by the horizontal line connecting σ_d with point A and the vertical line connecting point A with the original linear σ_θ curve. It is more likely that the actual shape follows the dashed curve after A, which eventually asymptotes to the linear solution. Irrespective of the model, the areas under the two curves, the linear solution and the modified distribution must be the same. For example, for the case of the second approximation, this means that the two shaded areas in Figure 2.11 must be equal.

In view of this reasoning, Equation 2.9 can be thought of as a statement of this equality of areas, which is a statement that the total force per unit width caused by the linear stress distribution equals the force per unit width caused by the modified distribution

$$\int_R^{R+\ell} \sigma_x(x = 0, y)\mathrm{d}y = \sigma_d\ell \tag{2.12}$$

Equation 2.12 is valid for any damage size ℓ. The left-hand side is the area under the linear curve, and the right-hand side is the area under the modified curve consisting of a horizontal segment and a vertical segment intersecting the linear curve at some distance ℓ from the hole edge. At failure, $\sigma_d = \sigma_f$ and $\ell = a_o$ and Equation 2.9 is recovered. The average stress in the damaged region (assuming the damaged region extends to a_o) equals the undamaged failure strength of the laminate.

As already mentioned, d_o and a_o are different. They are determined experimentally, and were originally assumed to be material constants. However, it was found that they change with the laminate and are therefore laminate constants. The variation from laminate to laminate is not drastic, and the Whitney–Nuismer approach can still be used to predict OHT or OHC values reasonably well. For typical carbon/epoxy materials, d_o is found to be 1.01 mm and a_o 3.8 mm. Predictions obtained using Equation 2.9 are compared with test results from [20] in Figure 2.12. The material used is AS1/3501-6, and d_o was set equal to 3.8 mm. The experimentally measured undamaged failure strength σ_f in the right-hand side of Equation 2.8 was obtained from [20].

The results in Figure 2.12 show that the Whitney–Nuismer model can give very good predictions if the value of a_o is known (3.8 mm was used here). Still, for the [30/−30/0]s laminate, the predictions are above the test results for hole diameters greater than ~8 mm. In addition, for other laminates, the predictions can be even worse. To use either of the Whitney–Nuismer criteria presented in this section, one must make sure that a good characteristic distance d_o or averaging distance a_o is used.

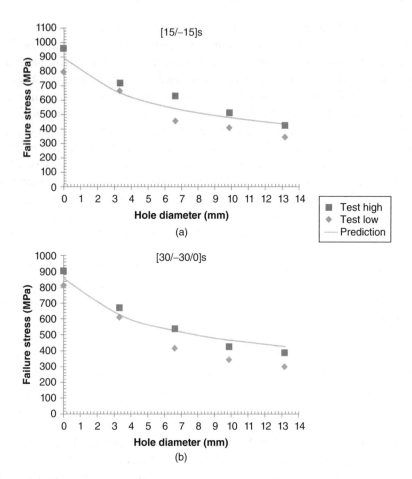

Figure 2.12   (a,b) Whitney–Nuismer average stress predictions versus test results

Or, it can be modified further, as will be shown in Section 2.6 so that the averaging distance can be obtained analytically.

2.5   Other Approaches to Predict Failure of a Laminate with a Hole

The results presented in the previous section indicate that, for improved accuracy in failure predictions, improved approaches may be necessary. This section discusses some of the relevant attempts.

   Generalised approaches for bolted joints can be used to recover the open-hole solution as a special case. One such approach was proposed by Garbo and Ogonowski [21]. In their general formulation, they model an infinite plate with a hole using complex elasticity. The fastener hole is assumed to have a sinusoidal stress distribution over half of its surface, corresponding to a bolt load. At the far field, any combination of in-plane loads can be applied. By setting the stress distribution corresponding to the fastener hole equal to zero, the case of an open hole is recovered.

Treating the plate as a homogeneous orthotropic plate, the in-plane stresses σ_x, σ_y and τ_{xy} at any point can be obtained. The problem is, again, how to use these stresses in a reliable failure criterion. Garbo and Ogonowski proposed an approach similar to that by Whitney and Nuismer. They used the classical laminated-plate theory to determine the individual ply stresses and evaluated them at a characteristic distance R_c away from the edge of the hole. They then used a maximum stress or strain criterion to predict failure. Unlike the Whitney–Nuismer method where laminate stresses are evaluated at a characteristic distance, here ply stresses are evaluated. The method works well and gives accurate predictions provided good test data are available to obtain R_c and first ply failures that do not lead to final failure are discarded. The latter requires some post-first-ply failure criterion and/or progressive failure analysis.

This approach has significant advantages: It can deal with any in-plane loading, and it performs the analysis on the ply level. Unfortunately, the characteristic distance R_c is still not a material constant. In addition, a good progressive failure analysis is needed to obtain accurate predictions for final failure.

Other methods have been proposed by Pipes *et al.* [22] and Soutis *et al.* [23]. In the first case, a three-parameter model is used to plot notched strength for all materials with the same SCF on the same master curve. It is based on curve-fitting, but can work well if the model parameters are determined accurately. In the second case, a linear elastic fracture mechanics approach is used. Damage is modelled as a crack emanating from the edge of the hole with a constant stress on its surface. This crack surface stress makes the model a one-parameter model. There are some questions about the use of classical fracture mechanics with stress singularity of 0.5, which will be discussed at some length in the next chapter. Despite these drawbacks the model gives good predictions provided the crack surface stress is obtained from test data.

All the methods discussed above try to avoid detailed modelling of the damage region and how it evolves as the load is increased. As a result, they cannot work in all cases and rely on parameters obtained by matching tests results. More elaborate methods are necessary that can model damage creation and evolution. A very promising attempt in that direction is the work by Maimí *et al.* [24, 25]. However, it is computationally intensive and requires the determination of several material constants, making it hard to use in day-to-day design work.

2.6 Improved Whitney–Nuismer Approach

The simplicity of the Whitney–Nuismer approach makes it a desirable tool to use if its accuracy can be improved. An approach to do so is given in this section. The approach aims at determining analytically the averaging distance a_o over which to average the linear stress solution from [10] and [11]. This distance will be a laminate constant.

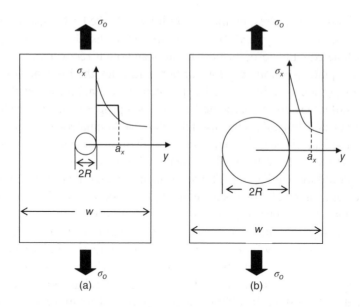

Figure 2.13 (a,b) Finite width specimens with holes under tension

If a_o is a laminate constant, then it will be the same for small and large holes, or it will be the same for all values of $2R/w$, where R is the hole radius and w the specimen width. This, of course, would be true as long as the ligament, that is, the amount of material between the hole circumference and the hole edge, which is equal to $w/2 - R$, is greater than a_o. The situation is shown in Figure 2.13.

Therefore, if a solution can be obtained for the failure stress for one of these geometries for which

$$\frac{w}{2} - R > a_o \tag{2.13}$$

then the value of a_o determined for that case will be valid for any other $2R/w$ ratio as long as Equation 2.13 is satisfied.

An approximate solution for failure of notched orthotropic laminates of finite width has been obtained by Tan [16]. In his solution, the far-field stress σ_o required to cause failure for a notch-sensitive material (see Section 1.2) is given by

$$\sigma_o = \frac{F_{tu}}{2k_t^\infty}\left[2 - \left(\frac{2R}{w}\right)^2 - \left(\frac{2R}{w}\right)^4 + \left(\frac{2R}{w}\right)^6 (k_t^\infty - 3)\left(1 - \left(\frac{2R}{w}\right)^2\right)\right] \tag{2.14}$$

where F_{tu} is the un-notched tension strength of the laminate.

In the limiting case when Equation 2.13 becomes an equality, that is, when the amount of material on either side of the hole is just equal to the averaging distance a_o,

$$\frac{w}{2R} - 1 = \frac{a_o}{R}$$

which can be rearranged as

$$\frac{2R}{w} = \frac{1}{1 + (a_o/R)} \tag{2.15}$$

Denoting, for simplicity, a_o/R as s, we have

$$\frac{2R}{w} = \frac{1}{1 + s} \tag{2.16}$$

This can be substituted in Equation 2.14, which can be rearranged to read

$$\frac{\sigma_o}{F_{tu}} = \frac{1}{2k_t^\infty} \left[2 - \frac{1}{(1 + s)^2} - \frac{1}{(1 + s)^4} + \frac{(k_t^\infty - 3)s(2 + s)}{(1 + s)^8} \right] \tag{2.17}$$

Equation 2.17 gives the far-field stress that would cause failure in a finite-width laminate with a hole as a function of the ratio a_o/R when the hole diameter and specimen width satisfy Equation 2.15. This is a special case and corresponds to the largest diameter that the hole can have without extending into the averaging distance a_o. This means that, of all $2R/w$ ratios, we select the one for which Equation 2.15 is satisfied, and thus the amount of material left outside the hole is exactly equal to the averaging distance a_o.

The stress σ_x at $x = 0$ (see Figure 2.13) is now determined as a function of y for a finite-width laminate. This is done by taking the stress for an infinite plate given by Equation 2.7 and modifying it to correspond to a finite-width plate. Equation 2.7 has two important properties: (i) when integrated from $y = R$ to infinity, it equals σ_o the applied stress, and (ii) when evaluated at $y = R$, it equals $k_t^\infty \sigma_o$, that is, it recovers the SCF for this laminate. The same two basic properties must be reproduced for a finite-width laminate but appropriately adjusted to reflect the fact that the laminate is no longer infinite. This means that (i) when integrated from $y = R$ to $w/2$, the net section stress $\sigma_o/(1 - 2R/w)$ must be recovered, which guarantees that the force along the line $x = 0$ equals half the total applied force for each side of the hole, and (ii) when evaluated at $y = R$, the SCF adjusted for finite width must be recovered. So, in a manner analogous to Equation 2.7, the following expression is assumed for a finite width plate:

$$\sigma_x(x = 0, y) = \sigma_o \left[1 + \frac{1}{2} \left(\frac{R}{y} \right)^2 + \frac{3}{2} \left(\frac{R}{y} \right)^4 + A_6 \left(\frac{R}{y} \right)^6 + A_8 \left(\frac{R}{y} \right)^8 \right] \tag{2.18}$$

Note that the first three terms in the square brackets on the right-hand side of Equation 2.18 are the same as in Equation 2.7. These correspond to the isotropic plate solution. The last two terms have the same R/y dependence as in Equation 2.7 but now the coefficients A_6 and A_8 are unknown. These two unknowns are determined by satisfying the two requirements mentioned above.

To satisfy the first requirement, the average value of σ_x from R to $w/2$ must be determined:

$$\sigma_{avg} = \frac{1}{(w/2) - R} \int_R^{w/2} \sigma_x(x = 0, y)dy$$

$$= \frac{\sigma_o}{(w/2) - R} \left[\begin{array}{l} \frac{w}{2} - R - \frac{R^2}{2}\left(\frac{1}{w/2} - \frac{1}{R}\right) - \frac{R^4}{2}\left(\frac{1}{(w/2)^3} - \frac{1}{R^3}\right) \\ \\ - \frac{A_6 R^6}{5}\left(\frac{1}{(w/2)^5} - \frac{1}{R^5}\right) - \frac{A_8 R^8}{7}\left(\frac{1}{(w/2)^7} - \frac{1}{R^7}\right) \end{array} \right] \quad (2.19)$$

Equating the right-hand side of Equation 2.19 to $\sigma_o/(1 - 2R/w)$ and rearranging yields

$$1 + \frac{1}{2}\left(\frac{2R}{w} - 1\right) + \frac{1}{2}\left(\left(\frac{2R}{w}\right)^3 - 1\right) + \frac{A_6}{5}\left(\left(\frac{2R}{w}\right)^5 - 1\right) + \frac{A_8}{7}\left(\left(\frac{2R}{w}\right)^7 - 1\right) = 0$$
$$(2.20)$$

The second requirement takes the form

$$\sigma_o k_t^{FW} = \sigma_o \left(1 + \frac{1}{2} + \frac{3}{2} + A_6 + A_8\right) \Rightarrow 3 + A_6 + A_8 = k_t^{FW} \quad (2.21)$$

where k_t^{FW} is the SCF for a plate with a hole of finite width and is given by [16]

$$k_t^{FW} = k_t^\infty \left[\frac{2}{2 - (2R/w)^2 - (2R/w)^4 + (2R/w)^6(k_t^\infty - 3)(1 - (2R/w)^2)} \right] \quad (2.22)$$

Note that this finite width correction factor is somewhat different from the one given by Equation 2.5. The one in Equation 2.5 is slightly less accurate than the present one (see also Exercise 1.3).

Equations 2.20 and 2.21, with the insertion of Equation 2.22, form a system of two equations in the two unknowns A_6 and A_8. With A_6 and A_8 determined, and assuming that Equation 2.15 is valid, Equation 2.19 can be rewritten as

$$\sigma_{avg} = \frac{\sigma_o}{s} \left[\begin{array}{l} s + \frac{1}{2(1+s)} - \frac{1}{2}\left(\frac{1}{(1+s)^3} - 1\right) \\ \\ - \frac{A_6}{5}\left(\frac{1}{(1+s)^5} - 1\right) - \frac{A_8}{7}\left(\frac{1}{(1+s)^7} - 1\right) \end{array} \right] \quad (2.23)$$

According to the Whitney–Nuismer model, at failure, the stress averaged over a characteristic distance a_o next to the hole equals the un-notched strength F_{tu}. At failure, therefore, Equation 2.23 reads

$$F_{tu} = \frac{\sigma_o}{s} \left[\begin{array}{l} s + \frac{1}{2(1+s)} - \frac{1}{2}\left(\frac{1}{(1+s)^3} - 1\right) \\ \\ - \frac{A_6}{5}\left(\frac{1}{(1+s)^5} - 1\right) - \frac{A_8}{7}\left(\frac{1}{(1+s)^7} - 1\right) \end{array} \right] \quad (2.24)$$

Using now Equation 2.17 to substitute in Equation 2.24, and rearranging gives

$$
\frac{2k_t^\infty}{2 - (1/(1+s)^2) - (1/(1+s)^4) + (k_t^\infty - 3)s(2+s)/(1+s)^8}
$$

$$
= \frac{1}{s} \left[\begin{array}{l} s + \dfrac{1}{2(1+s)} - \dfrac{1}{2}\left(\dfrac{1}{(1+s)^3} - 1\right) \\[2mm] - \dfrac{A_6}{5}\left(\dfrac{1}{(1+s)^5} - 1\right) - \dfrac{A_8}{7}\left(\dfrac{1}{(1+s)^7} - 1\right) \end{array} \right] \tag{2.25}
$$

which has only one unknown, $s = a_o/R$. With s known from Equation 2.25, the averaging distance a_o can be determined.

The procedure is, therefore, as follows:

1. Select specimen lay-up (symmetric and balanced), hole radius R and specimen width w.
2. Determine the SCF for the infinite plate k_t^∞ from Equation 2.1. If the value of θ which maximises the right-hand side of Equation 2.1 is not 90°, then the procedure given here must be modified to use the σ_θ stress at that critical location.
3. Select a value of $s = a_o/R$.
4. Pick a value of $2R/w$ and calculate the stress concentration factor k_t^{FW} from Equation 2.22.
5. Solve Equations 2.20 and 2.21 for A_6 and A_8.
6. Check if Equation 2.25 is satisfied. If not, return to step (3) and repeat the process for a new $2R/w$. If yes, check if Equation 2.16 is satisfied. If not, return to step (2) and repeat the process with a new value of s. If yes, the current values of a_o/R and $2R/w$ form the sought-for solution.
7. Given the value of R from step (1) and the value of a_o/R from the previous step, determine the averaging distance a_o. This value will be constant for this laminate for any hole radius (but constant width).
8. For any given specimen for which a strength prediction is needed, obtain $2R/w$ and calculate a_o/R with a_o from the previous step.
9. Determine k_t^{FW} from Equation 2.22, and A_6 and A_8 from solving Equations 2.20 and 2.21. These are different from those in steps (4) and (5) before because they correspond to the actual geometry for which the strength prediction is needed.
10. Determine the current $s = a_o/R$ and solve Equation 2.24 for σ_o/F_{tu}:

$$
\frac{\sigma_o}{F_{tu}} = \frac{s}{\left[\begin{array}{l} s + (1/2(1+s)) - (1/2)((1/(1+s)^3) - 1) - (A_6/5) \\ \cdot ((1/(1+s)^5) - 1) - (A_8/7)((1/(1+s)^7) - 1) \end{array} \right]} \tag{2.24a}
$$

11. Use F_{tu} for the current laminate to obtain the far-field stress σ_o to cause failure.

This process gives the ratio a_o/R as a laminate constant dependent only on the lay-up via the SCFs for infinite and finite plates. Its great advantage is that it does not rely on any test results to determine or fit the averaging distance. The main assumption that must be satisfied is that, for large values of $2R/w$, the far-field failure stress follows the curve described by Equation 2.14. Once a_o is determined, it can also be used for smaller $2R/w$ values for which Equation 2.14 is no longer valid. The extent to which this assumption is valid will be clearer in the following, where results of the present prediction method will be compared to test results.

As an additional useful result the value of a_o for different laminates is presented immediately below. Since a_o depends only on k_t^∞ and the specimen width, the results shown in Figure 2.14 are valid for any orthotropic material. The curve in Figure 2.14 was obtained by picking different values for k_t^∞ and following steps (1)–(11) in the procedure just described. The specimen width used in Figure 2.14 was 50 mm.

In reviewing Figure 2.14, it is interesting to note that, if highly orthotropic laminates with SCF greater than 6 are excluded, then in the range $2 < k_t^\infty < 6$ the average value of a_o is ~3.8 mm, which is the value mentioned in Section 2.4 and the value often used in practice on the basis of fitting experimental data.

The method presented in this section was used to compare failure predictions to test results for specimens with holes. The test results are from [20] on AS1/3501-6 material. The specimen width in all tests was 50 mm. Hole diameters ranged from 1.5 to 13.5 mm, which cover a range of $2R/w$ values from 3% to 26%. Test specimens with 4, 6 and 8 plies (ply thickness 0.135 mm) were tested to failure under tension. Five replicates were used for each hole diameter and for the un-notched strength tests. For each laminate, the experimentally measured un-notched strength value was used for F_{tu} in Equation 2.24a.

The predictions of the improved Whitney–Nuismer method are compared to test results for laminates [15/−15]s and [30/−30/0]s in Figure 2.15. These are the same laminates as in Figure 2.12.

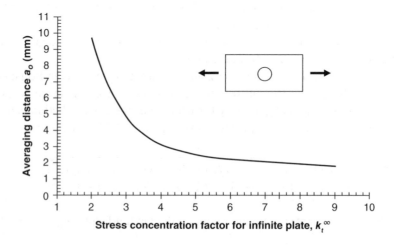

Figure 2.14 Averaging distance as a function of k_t^∞ for a 50-mm-wide specimen

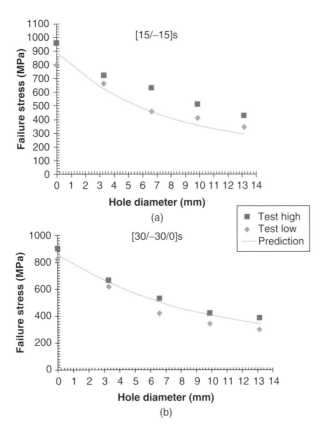

Figure 2.15 Improved Whitney–Nuismer predictions compared to test results: (a) [15/−15]s and (b) [30/−30/0]s laminates

These results compare quite well with the predictions in Figure 2.12, where a single value of a_o was used. In addition, now the case of [30/−30/0]s laminate is significantly improved compared to the original predictions in Figure 2.12. Predictions for two other laminates, namely [15/−15/0]s and [45/−45/0]s, are shown in Figure 2.16. Again, good agreement between the improved model and test results is observed.

As can be seen from Figures 2.15 and 2.16, the correlation between the improved model and test results is not perfect. In particular, the [15/−15/0]s laminate shows discrepancies for larger hole diameters. In fact, an extreme case where the improved model is not even close to test results is the [90]₄ laminate shown in Figure 2.17. This helps bring up some very important points related to the type of failure and the failure location.

The test results in Figure 2.17 have very wide scatter, with the coefficient of variation approaching 68% for some hole diameters, suggesting that more than one factor is responsible for final failure and not all of them are present in each specimen. It is also interesting, in this respect, that the test data for the largest diameter hole are higher than most data for the two smaller hole diameters.

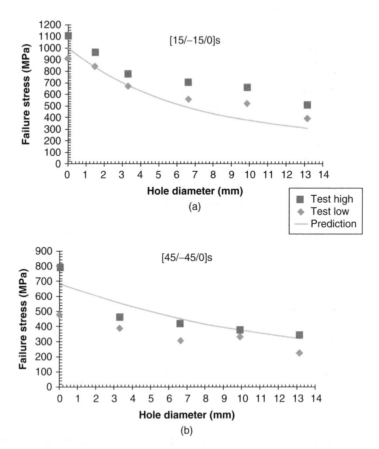

Figure 2.16 Improved Whitney–Nuismer predictions compared to test results: (a) [15/–15/0]s, (b) [45/−45/0]s laminates

From the perspective of failure mechanisms, the main ones are the transverse matrix cracks in the plies, the SCF caused by the hole and how the two interact. The effect of the matrix cracks is not captured by the present model. A refinement would adjust the stiffness properties for the laminate by reducing the stiffness transverse to the fibres for each ply and, to a lesser extent, the shear modulus of each ply. Then a new k_t^∞ would be computed and a new averaging distance. This effect is important because, if, for example, one assumed that as the load increases the transverse matrix cracks created (parallel to the fibres) reduce the transverse stiffness of the 90° plies to half their values, the failure location would move away from $\theta = 90°$. This is shown in Figure 2.18.

The reduced transverse stiffness case in Figure 2.18 has maximum circumferential stress (maximum stress concentration factor) occurring at $\theta = 65°$. This means that the approach presented in this section is no longer valid because σ_θ must be averaged at $\theta = 65°$ and not at $\theta = 90°$, which has been the premise in this section. In addition, as can be seen from Figure 2.18, for the reduced transverse stiffness, the maximum

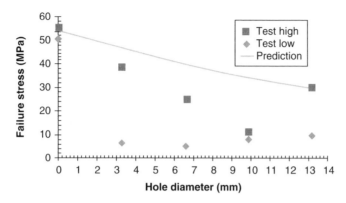

Figure 2.17 Improved Whitney–Nuismer predictions compared to test results: $[90]_4$ laminate

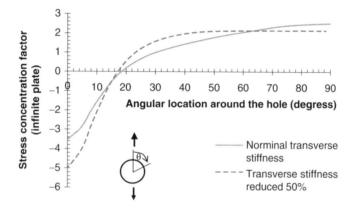

Figure 2.18 Stress concentration factor around the hole of a $[90]_4$ laminate

stress concentration factor is almost constant in a wide range between $\theta = 40°$ and $90°$. This suggests that local tiny defects away from the $\theta = 65°$ location might precipitate failure, introducing an element of randomness in the tests and perhaps explaining some of the wide scatter seen in Figure 2.17. Finally, the reduced transverse stiffness model in Figure 2.18 shows a significant increase in the compressive SCF from -3.5 to -5 at $\theta = 0°$. This suggests the possibility of compression-driven failure. For a compression-driven failure, the same approach as outlined in this chapter may be used but the averaging distance would be different as it would be associated with a different failure mode and corresponding strength. Alternatively, using the characteristic distance d_o instead of the averaging distance a_o, Kweon et al. [26] have shown that d_o for compression loading can be associated with the distance from the hole edge where the compression stress in a bolted joint is equal to the bearing stress $P/(Dt)$, where P is the applied load, D is the fastener diameter and t is the thickness of the fastened member in question.

This discussion on the $[90]_4$ laminate is meant to underline the fact that capturing the correct failure mode and its location are paramount in obtaining a good analytical model. Furthermore, other factors should not be overlooked, such as the presence of inter-laminar stresses at the hole edge, which, for some laminates, may drive failure.

2.7 Application: Finding the Stacking Sequence Which Results in Good OHT Performance

Two important competing factors were mentioned in the discussion of the results presented earlier in Table 2.1: (i) increasing the content of $0°$ plies in a laminate (with 0 direction aligned with the loading direction) increases the strength but also increases the SCF caused by the hole, and (ii) increasing the percentage of $45°$ plies reduces the strength but also reduces the stress concentration caused by the hole. In practice, laminates consisting of only $0°$ plies are not used because the loads are in multiple directions. Therefore, the question that arises is, what percentage of $0°$ and $45°$ plies should be used in a specific application for maximum load capability for a given thickness? Of course, without knowing the exact load cases, it is not possible to answer that question conclusively. However, some good starting designs can be found.

Two possibilities are considered. In the first, the laminate will consist of only $0°$ and $45°$ plies. In the second, the laminate will be required to obey the 10% rule, where there are at least 10% of the fibres in each of the four principal directions, 0, 45, −45 and 90. The improved Whitney–Nuismer approach of the previous section is used to find the best combination. A certain percentage of $45/−45°$ plies is selected, with the rest of the plies being 0 for the first case or 0 and 10% $90°$ plies for the second. The averaging distance a_o is then computed following the procedure in the previous section. This is then used to predict failure of a 50-mm-wide laminate with a 6.35-mm-diameter hole. The procedure is then repeated for different percentages of $45/−45$ plies. For the un-notched strength, the maximum stress criterion is used, which was found to be within 12% of the test results for the laminates tested. Material properties are the same as in [20].

The predicted OHT strength as a function of the percentage of $45/−45$ plies is shown in Figure 2.19. It is seen from Figure 2.19 that, if there are no $90°$ plies, the OHT strength drops as the percentage of $45/−45$ plies increases from 0% to about 20%. From 20% to 50% it stays relatively constant, and after 50% it decreases again. This 'plateau' between 20% and 50% is the range that should be used. In this range, the decrease in strength due to the increase in the percentage of $45°$ plies is offset by a corresponding reduction of the SCF caused by the hole. Values lower than 20% are not very useful because the laminate is not efficient under loading in directions other than the 0 direction. Values greater than 50% suffer from low strength because of a high percentage of $45/−45$ plies. Essentially, the strength reduction due to the increased content in $45/−45$ plies overwhelms the decrease in the SCF effect. One attractive feature of a relatively constant OHT strength in the range 20–50% is that one has the

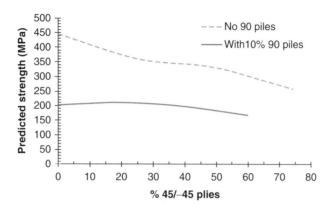

Figure 2.19 OHT strength for carbon/epoxy laminates consisting of 0/45/−45 plies or 0/45/−45 and 10% 90 plies

freedom to select the exact percentage of 45/−45 plies to suit the specific application taking into account additional load conditions as necessary. Essentially, any value in this range leads to approximately the same OHT strength, so the exact percentage of 45/−45 plies can now be selected based on other considerations.

A similar situation is observed with the lower curve in Figure 2.19 when 10% of the plies are 90° plies. In this case, there is a distinct maximum OHT strength occurring for ~25% 45/−45 plies. However, the variation of OHT strength around this maximum is small, and essentially one can again select any percentage of 45/−45 plies in the range 15–35% with little reduction in OHT strength. To be fair, any percentage of 45/−45 plies in the range 0–40% would have at most a 7% reduction from the maximum OHT strength (difference between the values at 40% and 25%). However, low percentages again are not usable in multi-load applications. So the best range is 20–40%, which is quite similar to that for the case where no 90 plies are present.

Exercises

2.1 The following material is given:

$E_x = 131$ GPa	$X_t = 2068$ MPa
$E_y = 11.51$ GPa	$X_c = 1724$ MPa
$v_{xy} = 0.29$	$Y_t = 103.4$ MPa
$G_{xy} = 4.83$ GPa	$Y_c = 330.9$ MPa
$t_{ply} = 0.1524$ mm	$S = 124.1$ MPa

On the basis of the stress concentration factor only, determine the value of φ in the laminate family $[\pm\varphi]s$ $(0 \le \varphi \le 90)$ such that the laminate is the strongest possible under tension.

2.2 Consider a square skin panel of side 0.12 m with lay-up [45/−45/0]s. Create a graph of the failure strength as a function of hole radius ranging from 5 mm to 5 cm using the Whitney–Nuismer method. Material properties are the same as in Exercise 2.1.

2.3 Refer to Figure E2.1 below. For the case of the hole with radius of 5 cm in the previous question, and assuming a notch-insensitive material, determine how much axial force goes through the portion that is taken up by the hole (as a fraction of the total load applied). Then, from Problem 2.2 calculate the 'real' stress concentration factor (failure stress without hole divided by failure stress with hole of radius 5 cm). If now there is no hole, determine the applied force at failure. Multiply the three numbers to obtain the load that the two edges of a flange around the hole at $\theta = 90$ would have to withstand so the laminate with a 5-cm radius hole fails at the same load as a laminate without a hole. If the flange lay-up is [45/−45/0]s^5 with the 0 in the circumferential direction, what is the flange height b? Why will this fail at $\theta = 0$ location (and thus this is not a good flange design)?

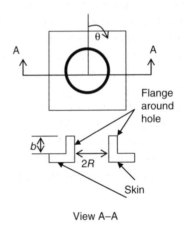

View A–A

Figure E2.1 Flanged hole for Exercise 2.3

2.4 The upper skin panels in an airplane wing are under compression (up-bending of the wing). At the same time, the lower skin panels are under tension with load equal in magnitude to the compressive load. Each panel consists of the skin between two adjacent stiffeners and two adjacent ribs and can be treated as simply-supported. The length a of each panel is 508 mm and the width b is 152.4 mm. The applied load, namely the tension for the lower panels and compression for the upper panels, is parallel to the dimension a of each panel.

To keep the cost down, the design under question uses exactly the same lay-up for upper and lower panels. It is a symmetric and balanced lay-up made up of a total of eight plies using only (some or all of) the orientations 0, 45, −45 and 90. The stress department is very proud of the design because the panels are optimised so they have the maximum possible load capacity and just fail at the critical load condition (obviously not all of them fail at the same time). Unfortunately, after the design was completed, the maintainability guys showed up and requested that circular cutouts be cut in each of the lower panels to be covered by access doors and to be used for inspection during service of the aircraft. Determine the largest possible diameter of the cutout in the lower panels if (a) it is assumed that the stress concentration obtained by analysis is correct and (b) using the Whitney–Nuismer approach (characteristic distance d_o only). Do not use any knock-downs.

The material used has the following properties:

E_x (Pa)	1.310E + 11	X_t (Pa)	2.068E + 09
E_y (Pa)	1.138E + 10	X_c (Pa)	1.379E + 09
v_{xy}	0.29	Y_t (Pa)	8.273E + 07
G_{xy} (Pa)	4.826E + 09	Y_c (Pa)	3.309E + 08
t_{ply} (mm)	0.1524	S (Pa)	1.241E + 08

2.5 For a $[0_8]$ laminate of the same material as in Exercise 2.4 and with the same panel width, determine the failure load N_x under tension with a hole of radius 33 mm using (a) the stress concentration factor and (b) the Whitney–Nuismer approach. What does this say about using the stress concentration factor approach?

2.6 State the condition for failure of a composite laminate with a hole of radius R and characteristic distance d_o (Whitney–Nuismer failure criterion). For the case of a quasi-isotropic lay-up, solve for d_o in closed form as a function of σ_{OH}/σ_{fu} the ratio of open hole to un-notched failure stress. Based on test data given in this chapter for quasi-isotropic lay-up, determine the value of $R/(R+d_o)$.

2.7 As R varies from 3.175 to 25.4 mm, obtain a plot of σ_{OH}/σ_{fu} for the three lay-ups: $[45/-45/0_4/90]s$ (fibre-dominated), $[45/-45/0/90]s$ (quasi-isotropic) and $[(45/-45)_2/0/90]s$ (matrix-dominated). Assume that d_o is the same as the value of d_o corresponding to quasi-isotropic lay-up with 6.35 mm diameter hole. Which lay-up is the strongest? What type of lay-up, namely quasi-isotropic, fibre-dominated or matrix-dominated, should therefore be preferred in the presence of holes? Discuss the implications of your results.

The material properties are as follows:

$$E_x = 137.9\,\text{GPa}$$
$$E_y = 11.7\,\text{GPa}$$
$$v_{xy} = 0.31$$
$$G_{xy} = 4.82\,\text{GPa}$$
$$t_{ply} = 0.1524\,\text{mm}$$

2.8 Create a plot of stress concentration factor as a function of location around a circular hole in a laminate under pure shear. The laminates to be considered are as follows:

[15/−15/0]s
[0]₆
[45/−45]s
Aluminium
For the composite, the following properties are given:
$$E_x = 130\,\text{GPa}$$
$$E_y = 10.5\,\text{GPa}$$
$$G_{xy} = 6.0\,\text{GPa}$$
$$v_{xy} = 0.28$$
$$t_{ply} = 0.1397\,\text{mm}$$

Which lay-up is best? How does the answer compare to the case where the laminate with the hole is under tension? (Figure E2.2).

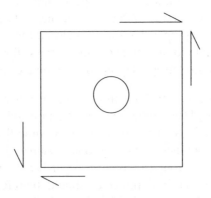

Figure E2.2 Loading situation for Exercise 2.8

References

[1] Portanova, M.A. and Masters, J.E. (1995) Standard Methods for Filled Hole Tension Testing of Textile Composites. AIAA Paper 96–0633 (1996), also NASA TM-107140.

[2] Chang, F. and Scott, R. (1982) Strength of mechanically fastened composite joints. *J. Compos. Mater.*, **16**, 470–494.

[3] Fan, W. and Qiu, C. (1993) Load distribution of multi-fastener laminated composite joints. *Int. J. Solids Struct.*, **30** (21), 3013–3023.

[4] Kradinov, V., Barut, A., Madenci, E. and Ambur, D. (1998) Bolted double-lap composite joints under mechanical and thermal loading. *Int. J. Solids Mech.*, **35** (15), 1793–1811.

[5] Hamada, H. and Maekawa, Z. (1996) Strength prediction of mechanically fastened quasi-isotropic carbon/epoxy joints. *J. Compos. Mater.*, **30**, 1596–1612.

[6] Hart-Smith, L.J. (1976) Bolted Joints in Graphite-Epoxy Composites. NASA CR-144899.

[7] Kradinov, V., Madenci, E. and Ambur, D.R. (2007) Application of genetic algorithm for optimum design of bolted composite lap joints. *Compos. Struct.*, **77**, 148–159.

[8] Hart-Smith, L.J. (1994) The key to designing efficient bolted composite joints. *Composites*, **25**, 835–837.

[9] Bakker, M. (2012) Analytical stress field and failure prediction of mechanically fastened composite joints. MS thesis. Delft University of Technology.

[10] Lekhnitskii, S.G. (1963) in *Theory of Elasticity of an Anisotropic Elastic Body* (Translated by P. Fern), Holden Day Inc., San Francisco, CA.

[11] Savin, G.N. (1961) in *Stress Concentration Around Holes* (Translated by W. Johnson), Pergamon Press.

[12] Lucking, W.M., Hoa, S.V. and Sankar, T.S. (1984) The effect of geometry on interlaminar stresses of [0/90]s composite laminates with circular holes. *J. Compos. Mater.*, **17**, 188–198.

[13] Carlsson, L. (1983) Interlaminar stresses at a hole in a composite member subjected to in-plane loading. *J. Compos. Mater.*, **17**, 238–249.

[14] Fagiano, C., Abdalla, M.M., Kassapoglou, C. and Gürdal, Z. (2010) Interlaminar stress recovery for three-dimensional finite elements. *Compos. Sci. Technol.*, **70**, 530–538.

[15] Saeger, K.J. (1989) An efficient semi-analytic method for the calculation of the interlaminar stresses around holes. PhD thesis. Massachusetts Institute of Technology.

[16] Tan, S.C. (1988) Finite width correction factors for anisotropic plate containing a central opening. *J. Compos. Mater.*, **22**, 1080–1097.

[17] Gürdal, Z. and Haftka, R. (1987) Compressive failure model for anisotropic plates with a cutout. *AIAA J.*, **11**, 1476–1481.

[18] Iarve, E.V., Mollenhauer, D., Whitney, T.J. and Kim, R. (2006) Strength prediction in composites with stress concentrations: classical weibull and critical failure volume methods with micromechanical considerations. *J. Mater. Sci.*, **41**, 6610–6621.

[19] Whitney, J.M. and Nuismer, R.J. (1974) Stress fracture criteria for laminated composites containing stress concentrations. *J. Compos. Mater.*, **8**, 253–265.

[20] Lagacé, P.A. (1982) Static tensile fracture of graphite/epoxy. PhD thesis. Massachusetts Institute of Technology.

[21] Garbo, S.P. and Ogonowski, J.M. (1981) Effect of Variances and Manufacturing Tolerances on the Design Strength and Life of Mechanically Fastened Composite Joints, AFWAL-TR-81-3041.

[22] Pipes, R.B., Wetherhold, R.C. and Gillespie, J.W. (1979) Notched strength of composite materials. *J. Compos. Mater.*, **13**, 148–160.

[23] Soutis, C., Fleck, N.A. and Smith, P.A. (1991) Failure prediction technique for compression loaded carbon fibre-epoxy laminate with open holes. *J. Compos. Mater.*, **25**, 1476–1498.

[24] Maimí, P., Camanho, P., Mayugo, J. and Dávila, C. (2007) A continuum damage model for composite laminates: part I. Constitutive model. *Mech. Mater.*, **39**, 897–908.

[25] Maimí, P., Camanho, P., Mayugo, J. and Dávila, C. (2007) A continuum damage model for composite laminates: part II. Computational implementation and validation. *Mech. Mater.*, **39**, 909–921.

[26] Kweon, J.H., Ahn, H.S. and Choi, J.H. (2004) A new method to determine the characteristic lengths of composite joints without testing. *Compos. Struct.*, **66**, 305–315.

3

Cracks

3.1 Introduction

In metals, damage, especially under cyclic loading, manifests itself as a crack. In composites, cracks appear only as a result of specific threats such as impact with sharp objects that manage to penetrate the entire laminate. Usually, damage in composites manifests itself as a combination of matrix cracks, delaminations and fibre breakage that does not form a well-defined through-the-thickness crack. As a result, cracks are relatively rare in composite structures but, nevertheless, quite important because of their effect on residual strength.

The bi-material nature of a composite laminate and the fact that it consists of layers with fibres in different directions do not promote the creation of a through-the-thickness crack. In addition, once such a crack is created, they do not allow the more 'classical' crack behaviour that is observed in metals. If the crack in a composite laminate grows under load, it usually does not do so in a self-similar manner. It follows a 'jagged' path defined and constrained by the fibre orientations in the different plies. In general, fibres act as crack stoppers. The matrix having low toughness allows the crack to grow easily until it reaches a fibre that slowdowns the crack or stops it altogether. This is the reason why matrix cracks in-between the fibres of a ply, for example, the 90° plies in a [0/90/0] laminate, are confined by the cross-fibres in adjacent plies. This is also the reason that cracks forming between plies, that is, delaminations, do not easily branch into the adjacent plies and grow mostly between plies. The fibres in the plies on either side of the ply interface act as crack stoppers.

Growth of through-thickness cracks in a composite is a multi-scale problem. At the scale of a laminate thickness, where the laminate is considered homogeneous, crack growth is, typically, not self-similar [1]. To be able to understand and predict such behaviour, it is necessary to go down to lower scales, accounting for the effects the matrix and fibres have on slowing down or promoting the crack growth. As in the case of holes in the previous chapter, accurate modelling of damage creation ahead of

Modeling the Effect of Damage in Composite Structures: Simplified Approaches, First Edition. Christos Kassapoglou.
© 2015 John Wiley & Sons, Ltd. Published 2015 by John Wiley & Sons, Ltd.

a crack tip requires very detailed models that may have to account for fibres and the matrix separately in order to capture the details of crack growth.

3.2   Modelling a Crack in a Composite Laminate

In general, fibres are tougher than the matrix. A crack grows more easily in the matrix than in a fibre. Even toughened matrices still have lower toughness than glass or carbon fibres. Therefore, a crack in a composite will exhibit one of the two behaviours shown in Figure 3.1. In the first case, a crack in the matrix grows relatively easily until it reaches the next fibre where it stops (or slows down significantly). In the second case, a crack inside a fibre again grows relatively easily and then more rapidly into the matrix until it reaches again another fibre. Therefore, the critical factor affecting the crack growth is the matrix/fibre interface for both cases of Figure 3.1. While the details of crack growth will, to a lesser extent, be affected by the characteristics of growth inside a fibre or in the matrix, the main contributor will be the matrix/fibre interface. A zoomed view of this situation is shown in Figure 3.2.

The problem of Figure 3.2 is that of a crack terminating at the interface between two dissimilar bodies. This problem was first solved by Fenner [2] and was later adapted for a composite material by Mar and Lin [3, 4].

The most important difference is shown in the right of Figure 3.2. Unlike metals where the stress σ_x near the crack tip is proportional to $1/y^{1/2}$, the σ_x stress in a bi-material interface is proportional to $1/y^m$. The stress is singular with stress singularity m, which, for typical composites, is less than 0.5. The strength of the singularity m depends on the individual stiffness properties of the fibre and the matrix.

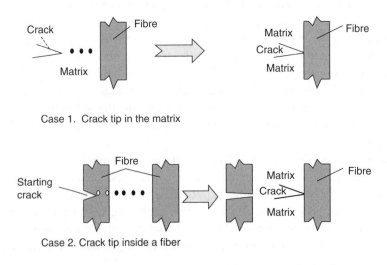

Case 1. Crack tip in the matrix

Case 2. Crack tip inside a fiber

Figure 3.1   Different cases of a crack in a composite laminate

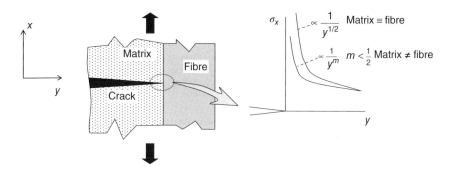

Figure 3.2   Crack at a matrix/fibre interface

The dominant stress term near a crack tip in a metal is given by

$$\sigma_x \propto \frac{K}{\sqrt{\pi}y^{1/2}}$$ (3.1)

where K is the stress intensity factor.

The corresponding stress expression for a composite with a crack at the matrix/fibre interface was proposed by Mar and Lin in the form:

$$\sigma_x \propto \frac{H}{y^m}$$ (3.2)

where H is analogous to the stress intensity factor and m is the strength of the singularity that is, typically, less than 0.5.

In view of Equation 3.2, Mar and Lin postulated that the failure strength σ_f in a composite laminate with a notch can be predicted by the relation:

$$\sigma_f = \frac{H_c}{(2a)^m}$$ (3.3)

In Equation 3.3, H_c is referred to as the *composite fracture toughness* and $2a$ is the notch size.

For the first-generation graphite/epoxy material (AS1/3501-6), Mar and Lin determined analytically the value of m to be 0.28. For boron/aluminium, they obtained experimentally a value (by fitting a curve through test results for different crack sizes) of 0.33. It is important to note that the value of m obtained by considering the idealisation of Figure 3.2 is independent of the laminate layup. However, test results show that there is some dependence on layup. For example, for a $[\theta/0/-\theta]$s AS1/3501-6 laminate, the value of m obtained experimentally as θ varies is in the range of 0.28–0.334. Somewhat larger variations from the nominal value of 0.28 are observed for $[\pm\theta]$s, $[0/\pm\theta]$s and $[\pm\theta/0]$s laminates.

This dependence of m on layup suggests that, for improved accuracy, Equation 3.3 must be treated as a two-parameter model where both H_c and m are obtained by

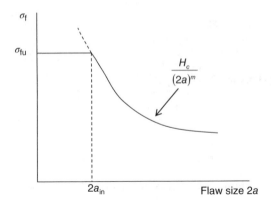

Figure 3.3   Failure strength of a notched laminate as a function of notch size

fitting this equation to experimental data. This limits the applicability of the Mar–Lin approach because of the need for test results to accurately determine the two parameters. Alternatively, one can use the analytically predicted value of m and only treat H_c as a fitting parameter, which, in some cases, reduces the accuracy (see below).

Equation 3.3 does not explicitly depend on the shape of the notch. It can equally be applied to holes and cracks. $2a$ can be the hole diameter or the crack size.

An interesting implication of Equation 3.3 is the existence of an inherent flaw size. As the flaw size $2a$ is zero, Equation 3.3 predicts the infinite strength. In reality, the plot of failure strength as a function of flaw size is truncated by the undamaged failure strength σ_{fu} as shown in Figure 3.3.

The curve described by Equation 3.3 intersects the undamaged failure strength at a flaw or crack size equal to $2a_{in}$. Then,

$$\sigma_{fu} = \frac{H_c}{(2a_{in})^m} \tag{3.4}$$

Solving for $2a_{in}$ gives

$$2a_{in} = \left(\frac{H_c}{\sigma_{fu}} \right)^{\frac{1}{m}} \tag{3.5}$$

Equations 3.4 and 3.5 imply that there is an inherent flaw size present in a composite laminate that precipitates un-notched or pristine failure. Substituting typical values for carbon epoxy gives a value of $2a_{in}$ of 0.13 mm. Then, if there is a hole or a crack of size less than 0.13 mm, the laminate behaves as if the hole or the crack is not there and fails at the undamaged failure strength.

It is interesting to note that this value of $2a_{in}$ corresponds, approximately, to 16 fibre diameters for carbon fibres with an approximate diameter of 8 μm. This gives an idea of the different scales that one would have to bridge in order to create an accurate analysis model: from one fibre diameter, 8 μm, to about 15–20 fibre diameters, which

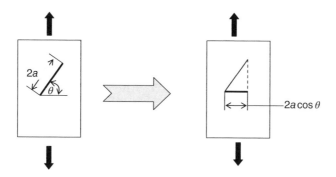

Figure 3.4 Analysis of cracks inclined to the loading direction

is close to a ply thickness of 0.13–0.2 mm, at which length scale smearing fibre and matrix properties is valid to the laminate thickness that ranges from a few millimetres to a few centimetres. For example, going from one fibre diameter to a laminate thickness of 4 mm corresponds to two to three orders of magnitude (see also Figure 1.1). Obviously, properties that are important at a low scale may not be important at a higher scale or, more likely, may combine to form parameters that govern the behaviour of higher scales. This is the challenge in modelling damage such as cracks in composite materials: to create models that cross the different scales accurately and, at the same time, are computationally efficient.

One advantage of the Mar–Lin approach is that it is independent of the shape of the notch. Equation 3.3 is valid, with appropriate H_c and m values, for both a hole and a crack of the same size $2a$. Also, if a crack is inclined at an angle, see Figure 3.4, to find the strength of the laminate under tension it suffices to use the projected crack length perpendicular to the direction of loading.

3.3 Finite-Width Effects

The discussion in the previous section was for infinite plates. The same as for holes, see Section 2.1, the analysis for cracks in composite laminates of finite width requires the use of a finite-width correction factor. It has been found that the finite-width correction factor for a centre crack in an isotropic plate under tension can be used for orthotropic plates:

$$\text{FWCF} = 1 + 0.1282\frac{2a}{w} - 0.2881\left(\frac{2a}{w}\right)^2 + 1.5254\left(\frac{2a}{w}\right)^3 \tag{3.6}$$

For orthotropic plates with $2a/w < 0.5$, Equation 3.6 is accurate to within 3%. Then, the failure prediction for a finite-width plate with a crack would be obtained by combining Equations 3.3 and 3.6:

$$\sigma_f = \frac{H_c}{(2a)^m(\text{FWCF})} \tag{3.7}$$

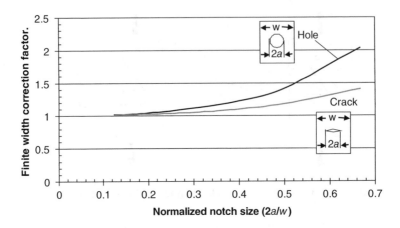

Figure 3.5   Finite-width effects for holes and cracks

It is interesting to compare the finite-width correction factor for a crack from Equation 3.6 to that for a hole from Equation 2.5. This is done in Figure 3.5 (see also [5]). As can be seen from Figure 3.5, the two correction factors are very close to each other up to a notch size of $2a/w \approx 0.3$. For larger notch sizes, the correction factor for holes is more severe resulting in a hole being more critical than a crack of the same size. Note that this conclusion is only for tension loading.

3.4   Other Approaches for Analysis of Cracks in Composites

The Mar–Lin approach, which was the focus of Section 3.2, suffers from the fact that it is essentially a two-parameter model. The fracture toughness H_c and the strength m of the stress singularity at the crack tip must be determined experimentally in order to achieve the highest possible accuracy. Therefore, improved analysis methods are needed. Over the years, several researchers proposed different methods.

Whitney and Nuismer [6] extended their approach for holes discussed in Section 2.4. Poe [7] and Poe and Suva [8] used a similar approach as Whitney and Nuismer, but instead of stress at a distance they used strain at a distance in their failure criterion. Efforts to model the damage and its evolution based mainly on finite elements have been proposed by Aronsson and Bäcklund [1] and Chang and Chang [9].

An excellent discussion of some of these methods was provided by Walker *et al.* [10]. In this work, test results obtained for different crack sizes' layups and materials were compared to different analytical predictions. For a fair comparison, to eliminate shifts in the predictions caused by the specifics of each method, all methods were forced to go through one test datum point at a specific crack length. The results of this comparison are shown in Figures 3.6–3.8.

The effect of crack sizes from 6.35 to 44.5 mm for a 16-ply AS4/938 layup is shown in Figure 3.6. The prediction methods used were Mar–Lin, Poe–Sova, Whitney–Nuismer and Linear Elastic Fracture Mechanics (LEFMs). All methods

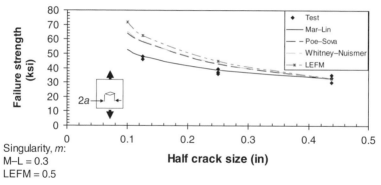

Singularity, *m*:
M–L = 0.3
LEFM = 0.5

All prediction methods are forced to go through the average test
result at 2*a* = 0.875 in

Figure 3.6   Test versus predictions for a 16-ply AS4/938 CFRP layup with a centre crack

were forced to match test results at a crack length of 44.5 mm. For Mar–Lin, the
stress singularity was set to 0.3 (which is very close to the analytically determined
value of 0.28 for the similar material), and for LEFM, the strength of the singularity
is the classical value of 0.5. It can be seen from Figure 3.6 that the Mar–Lin approach
gives excellent predictions. The remaining methods are un-conservative with LEFM
giving the worst predictions. As might be expected, because they are both stress and
strain at a point method, the Poe–Suva and the Whitney–Nuismer methods are very
close to each other but un-conservative compared to the test results.

The situation for a 10-ply laminate of the same material is shown in Figure 3.7. The
range of crack sizes covered is 6.35–127 mm. Here, all prediction methods are forced
to match test results at a crack length of 127 mm. Again, the Mar–Lin predictions with
$m = 0.3$ are the best but are slightly un-conservative for small crack sizes. The other
three methods give worse predictions with LEFM again being the worst.

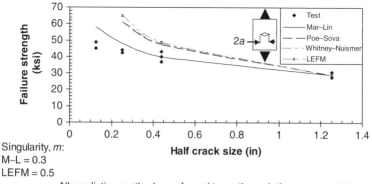

Singularity, *m*:
M–L = 0.3
LEFM = 0.5

All prediction methods are forced to go through the average test
result at 2*a* = 2.5 in

Figure 3.7   Test versus predictions for a 10-ply AS4/938 CFRP layup with a centre crack

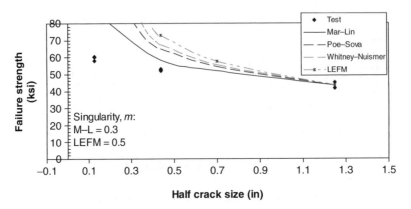

All prediction methods are forced to go through the average test result at
2a = 2.5 in

Figure 3.8   Test versus predictions for a 25%T100/75%AS4/938 layup with a centre crack

The effect of mixing two different carbon fibres, T100 and AS4, in the same laminate is examined in Figure 3.8. The range of crack sizes is the same as in the previous figure, 6.35–127 mm. All methods are forced to match the average test result
at 127 mm. In this case, all methods are un-conservative but, again, the Mar–Lin
approach with $m = 0.3$ is the closest to test results. As before, LEFM gives, again,
the worst predictions. The presence of two different fibres clearly renders the use of
$m = 0.3$ for Mar–Lin or $m = 0.5$ for LEFM inadequate. It is expected that the different fibres affect the tendency of the crack to grow in a way that cannot be captured
by a single exponent in the stress expression. Similarly, smearing the properties in the
Poe–Suva and Whitney–Nuismer methods is also not accurate enough. More elaborate
methods are needed.

Two main conclusions can be drawn from the above discussion. One, the Mar–Lin
approach gives the best predictions but is not always satisfactory and modifications or
improved methods are necessary. Two, using the classical singularity strength of 0.5
in composite laminates with cracks is not a good idea as it gives the worst predictions
among the methods compared.

As already mentioned, more accurate models require computer simulation that
allows crack creation and growth in any direction, independent of a finite element
mesh for example. This means that the simulation must be able to deal with
discontinuities such as cracks. One way this can be achieved is by enriching the
approximation or interpolation functions used in a finite element formulation. This,
typically, requires the use of additional degrees of freedom that allow modelling any
discontinuities that a crack would cause but do not require explicit modelling of the
crack in the finite element mesh. In this respect, eXtended Finite Element Method
(XFEM) is a very promising approach [11].

3.5   Matrix Cracks

A special case of cracks that do not necessarily extend all the way through the thickness of a laminate is a matrix crack, see Figure 3.9. Matrix cracks are typically confined within plies of the same orientation and terminate at the interface with plies of a different orientation. They are created when the stress perpendicular to the fibres in a ply exceeds the transverse tension strength of that ply. This transverse strength changes with the thickness of the ply and whether the ply is wholly contained between plies of different orientation or it has one free surface at the top or bottom of a laminate. For this reason, researchers [12] have introduced the term *in situ* strength to refer to the varying strength of a ply perpendicular to the fibres. It has been shown [13] that as the thickness of the ply increases, the *in situ* strength decreases and approaches the transverse strength of a laminate consisting exclusively of 90°plies.

The onset of matrix cracking can be predicted using fracture mechanics. This requires invoking an assumption that the lower scale damage in the form of tiny matrix cracks, voids and inconsistencies in matrix content, which are responsible for crack nucleation, can be represented by a longer scale crack. This avoids the complexity of having to model the lower scale phenomena. Dvorak and Laws [13] have shown that the *in situ* transverse tension strength Y_{is}^t of an embedded ply is

$$Y_{is}^t = 1.58Y^t \tag{3.8a}$$

when the ply is thick and

$$Y_{is}^t = \sqrt{\dfrac{4G_{IC}}{\pi t \left(\dfrac{1}{E_{22}} - \dfrac{v_{21}^2}{E_{11}} \right)}} \tag{3.8b}$$

when the ply is thin.

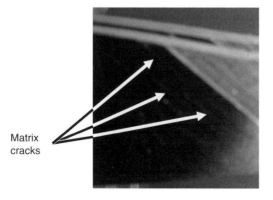

Matrix
cracks

Figure 3.9   Matrix cracks along −45° plies in carbon/epoxy specimen (*See insert for colour representation of this figure.*)

In Equations 3.8a and 3.8b, Y^t is the (unconstrained) transverse strength of the material, G_{IC} is the experimentally measured critical energy release rate in mode I for a crack in a 90° ply growing perpendicular to a tensile load transverse to the fibres in the 90°ply, t is the thickness of the ply and E_{11}, E_{22} and v_{21} are the Young's moduli parallel and perpendicular to the fibres and the minor Poisson's ratio, respectively. Usually, one uses Equation 3.8b unless it is shown that, for the t value of interest, the right-hand side of Equation 3.8b has already approached the right-hand side of Equation 3.8a.

Once a single crack is present, subsequent cracks can be predicted by calculating the transverse stresses in the vicinity of the matrix crack, parallel to the applied load, and using Equation 3.8 or applying a stress failure criterion [14]. Failure criteria that combine in-plane and out-of-plane stresses are more accurate as they account for all stresses present.

Consider as an example the situation of a $[0_n/90_m]$s laminate loaded in tension along the 0° fibres. This is shown in Figure 3.10.

If the laminate is long in the y direction, there is no dependence on the coordinate y and the equilibrium equations in each ply can be written as

$$\frac{\partial \sigma_x}{\partial x} + \frac{\partial \tau_{xz}}{\partial z} = 0$$

$$\frac{\partial \tau_{xz}}{\partial x} + \frac{\partial \sigma_z}{\partial z} = 0$$

(3.9a–b)

Two coordinate systems are set up as shown in Figure 3.5, one for the 0° and one for the 90° plies.

Assume now that the σ_x stress in the 90° and 0° plies, respectively, is given by

$$\sigma_{x1} = \sigma_{xff1} + A_1 f(x)$$

$$\sigma_{x2} = \sigma_{xff2} + A_2 f(x)$$

(3.10a–b)

where the subscript 1 denotes 90° plies and the subscript 2 refers to 0° plies.

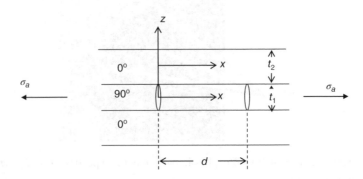

Figure 3.10   Matrix cracks in the 90° plies of a $[0_n/90_m]$s under tension

The far-field stresses σ_{xffi} in Equations 3.10a–b are obtained from classical laminated-plate theory as

$$\sigma_{xff1} = \frac{E_{22}(t_1 + 2t_2)}{E_{22}t_1 + 2E_{11}t_2}\sigma_a$$

$$\sigma_{xff2} = \frac{E_{11}(t_1 + 2t_2)}{E_{22}t_1 + 2E_{11}t_2}\sigma_a \qquad (3.11a\text{–}b)$$

with E_{11} and E_{22} as defined before, t_1 and t_2 as shown in Figure 3.10 and σ_a the far-field stress applied to the laminate.

Force equilibrium can be used to determine the constants A_1 and A_2. The sum of the axial forces in the 0° and 90° plies, away from the first crack, must add to the product of the applied stress σ_a times the cross-sectional area of the laminate. Setting A_1 equal to 1 without loss of generality, it can be shown that

$$A_2 = -\frac{t_1}{2t_2} \qquad (3.12)$$

The equilibrium equations (3.9a–b) are then used to determine the general expressions for the interlaminar stresses σ_z and τ_{xz}. The boundary conditions that σ_z and τ_{xz} be zero at the top (and bottom) of the laminate are satisfied next along with the stress continuity condition that σ_z and τ_{xz} be continuous at the 0/90 ply interface. The resulting stress expressions are

$$\sigma_{x1} = \sigma_{xff}^{(90)} + f(x)$$

$$\tau_{xz1} = -zf'(x)$$

$$\sigma_{z1} = \left(-\frac{t_1 t_2}{4} - \frac{t_1^2}{8} + \frac{z^2}{2}f''(x)\right)$$

$$\sigma_{x2} = \sigma_{xff}^{(0)} - \frac{t_1}{2t_2}f(x)$$

$$\tau_{xz2} = \left(-\frac{t_1}{4} + \frac{t_1}{2t_2}z\right)f'(x)$$

$$\sigma_{z2} = \left(-\frac{t_1 t_2}{16} + t_1\frac{z}{4} - \frac{t_1 z^2}{4t_2}\right)f''(x) \qquad (3.13)$$

The stress σ_y, not present in the equilibrium equations (3.9a–b), is determined by using the inverted stress–strain relations and strain compatibility [15]. The result is

$$\sigma_{y1} = (k_0)_1 + (k_1)_1 z - \left(\frac{S_{12}}{S_{22}}\right)_1 \sigma_x - \left(\frac{S_{23}}{S_{22}}\right)_1 \sigma_z$$

$$\sigma_{y2} = (k_0)_2 + (k_1)_2 z - \left(\frac{S_{12}}{S_{22}}\right)_2 \sigma_x - \left(\frac{S_{23}}{S_{22}}\right)_2 \sigma_z \qquad (3.13a\text{–}b)$$

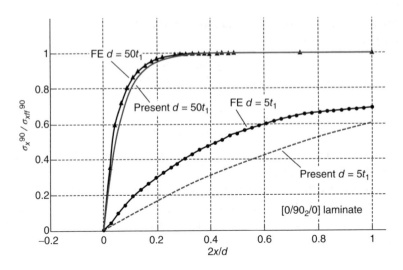

Figure 3.11   Normalised axial stress in 90° plies compared to finite element results

where, again, the subscripts 1 and 2 refer to the 90 and 0 plies, respectively. S_{ij} are ply compliances [16] and k_0 and k_1 are constants that recover the classical laminated-plate theory solution away from the crack.

The function $f(x)$ in Equations 3.13 is still unknown. This is determined by minimising the laminate energy using variational calculus [17]. This leads to a fourth-order ordinary differential equation with constant coefficients. Its solution is in terms of exponentials in x.

Stress solutions obtained with the present approach are compared to finite element results from [18] in Figure 3.11 for two different crack spacings d, see Figure 3.10. For the large crack spacing, the present solution is in excellent agreement with finite element results.

As the crack spacing becomes smaller, the present solution departs from the finite element results. This is mainly due to the fact that, in the present solution, the σ_x stress is independent of z in each ply.

Matrix cracks and their behaviour as predicted by this solution will be discussed later in some detail in Section 6.6.2 where fatigue of cross-ply laminates is briefly presented.

Exercises

3.1   Use the Whitney–Nuismer failure criterion for a hole and the Mar–Lin failure criterion for a crack to obtain a design chart that relates the crack size to the hole diameter with the same failure strength. This is an attempt to see whether a single curve can cover 'all' laminates. If this works then you can model a crack as a hole of equivalent diameter because the analysis is easier and does not require knowledge of H_c and m. Assume that your structure is a fuselage skin between two adjacent frames and stiffeners. The skin thickness is approximately 3.6–3.7 mm and its layup can be any of the following: 25/50/25, 16/68/16, 50/33/17, where the numbers denote percentages of 0°, +45°/−45° and 90° plies, respectively. Note that the 0° plies are in the hoop direction. The ply thickness is 0.1524 mm and you will have to round up or down if you cannot get exactly the percentages given. The graphite/epoxy material has basic ply properties:

$$E_x = 131\,\text{GPa}$$
$$E_y = 11.4\,\text{GPa}$$
$$v_{xy} = 0.31$$
$$G_{xy} = 5.17\,\text{GPa}$$
$$t_{ply} = 0.1524\,\text{mm}$$
$$X_t = 2068\,\text{MPa}$$
$$X_c = 1723\,\text{MPa}$$
$$Y_t = 68.9\,\text{MPa}$$
$$Y_c = 303.3\,\text{MPa}$$
$$S = 124.1\,\text{MPa}$$

The fuselage is under 1.3 atm pressure (overpressure case). It can be assumed to be a cylindrical pressure vessel under uniform internal pressure so the pressure load becomes a hoop stress and a longitudinal stress in the skin. For a crack parallel to the longitudinal direction as shown in Figure E3.1 and varying in length from 0% to 85% of the available width, determine the hole size (diameter) that has the same failure strength. Plot hole size as a function of crack size for the three layups given above (all in one chart). Assume that the longitudinal stress has no effect on a crack parallel to the longitudinal direction.

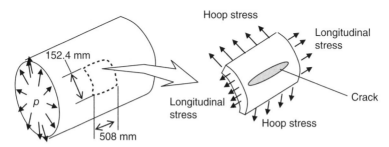

Figure E3.1 Pressurised fuselage skin with crack (Exercise 3.1)

The fuselage radius R is 3 m. For the strength analysis, you can assume that the skin is flat. The Mar–Lin constant for this material is $m = 0.28$ and the H_c values are, respectively, 453.9, 374.5 and 684.6 MPa mm$^{0.28}$.

(a) Obtain a chart of hole size versus crack size (for the same strength) for the three laminates. Discuss the relative sizes of equivalent crack and hole. What can you conclude? Can this be used?

(b) Determine the maximum crack and hole sizes that can be present in the centre of the panel without failure of the fuselage skin. What do you conclude about which type of laminate is best for this application/loading?

3.2 Testing a carbon/epoxy laminate with stacking sequence [0/30/−30]s has shown that the strength in the presence of cracks of various sizes follows the graph in Figure E3.2.

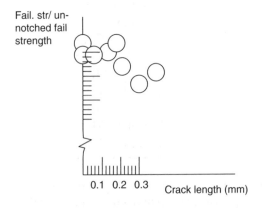

Figure E3.2   Test data for Exercise 3.2

Use the Mar–Lin approach to determine the failure strength (in megapascal) for a 15-mm long crack. Assume that $m = 0.3$ and material properties as follows:

E_x (Pa)	1.310E+11	X_t (Pa)	2.068E+09
E_y (Pa)	1.138E+10	X_c (Pa)	1.379E+09
v_{xy}	0.29	Y_t (Pa)	8.273E+07
G_{xy} (Pa)	4.826E+09	Y_c (Pa)	3.309E+08
t_{ply} (mm)	0.1524	S (Pa)	1.241E+08

3.3 The graph in Figure E3.3 (from Drury and Watson 'Good Practices in Visual Inspection', 2002) shows the probability of detection of a surface crack (e.g. on a fuselage or wing skin) as a function of crack size.

For a standard graphite/epoxy material, the experimentally determined H_c value for a series of laminates is given in Table E3.1. Determine the strength knockdown for each of the laminates in the table that should be applied to the undamaged laminate in order to cover cracks with 90% probability of detection and 95% probability of detection. How does that compare to the typical knockdown of 65% used for impact damage sometimes in practice? Solve the problem for two different cases: (i) $m = 0.28$ for all laminates and (ii) $m =$ given value in table.

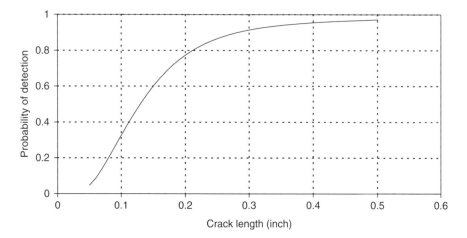

Figure E3.3 Probability of detection versus crack length (Exercise 3.3)

Table E3.1 Composite fracture toughness for various laminates

Laminate	M	H_c (MPa mm^m)
15A0	0.399	1123.696
30A0	0.419	788.392
45A0	0.188	182.484
60A0	0.249	94.0706
75A0	0.264	43.054
90A0	0.399	27.9741
15A1	0.284	1020.954
30A1	0.442	1095.255
45A1	0.233	568.764
60A1	0.259	748.954
75A1	0.242	737.856
90A1	0.276	886.774
15B1	0.293	1006.107
30B1	0.307	762.615
45B1	0.292	629.6
60B1	0.337	807.488
75B1	0.218	708.811
90B1	0.104	662.46

Note: $15A0 = [15/-15]s$; $15A1 = [15/-15/0]s$; and $15B1 = [0/15/-15]s$.

References

[1] Aronsson, C.G. and Bäcklund, J. (1986) Damage mechanics analysis of matrix effects in notched laminates, in *Composite Materials, Fatigue and Fracture*, American Society for Testing and Materials (ASTM), pp. 134–157, ASTM STP 907.

[2] Fenner, D.N. (1976) Stress singularities in composite materials with an arbitrarily oriented crack meeting an interface. *Int. J. Fract.*, **12**, 705–721.

[3] Mar, J.W. and Lin, K.Y. (1977) Fracture of boron/aluminum composites with discontinuities. *J. Compos. Mater.*, **11**, 405–421.

[4] Mar, J.W. and Lin, K.Y. (1977) Fracture mechanics correlation for tensile failure of filamentary composites with holes. *J. Aircr.*, **14**, 703–704.

[5] Walker, T.H., Avery, W.B., Ilcewicz, L.B. *et al.* (1991) Tension fracture of laminates for transport fuselage part 1: material screening. 2nd NASA Advanced Composites Technology Conference, 1991, pp. 197–238.

[6] Whitney, J.M. and Nuismer, R.J. (1974) Stress fracture criteria for laminated composites containing stress concentrations. *J. Compos. Mater.*, **8**, 253–265.

[7] Poe, C.C. Jr., (1983) A unifying strain criterion for fracture of fibrous composite laminates. *Eng. Fract. Mech.*, **17**, 153–171.

[8] Poe, C.C. Jr, and Sova, J.A. (1980) Fracture Toughness Of Boron/Aluminum Laminates with Various Proportions of 0° and ±45° Plies. NASA Technical Paper 1707.

[9] Chang, F.K. and Chang, K.Y. (1987) A progressive damage model for laminated composites containing stress concentrations. *J. Compos. Mater.*, **21**, 834–855.

[10] Walker, T.H., Avery, W.B., Ilcewicz, L.B. *et al.* (1991) Tension fracture of laminates for transport fuselage part 1: material screening. 2nd NASA Advanced Composites Technology Conference, 1991, pp. 197–238

[11] Mohammadi, S. (2012) *XFEM Fracture Analysis of Composites*, Chapter 4, John Wiley & Sons, Ltd, Chichester.

[12] Camanho, P.P., Dávila, C.G., Pinho, S.T. *et al.* (2006) Prediction of in situ strengths and matrix cracking in composites under transverse tension and in-plane shear. *Composites Part A*, **37**, 165–176.

[13] Dvorak, G.J. and Laws, N. (1987) Analysis of progressive matrix cracking in composite laminates II. First ply failure. *J. Compos. Mater.*, **21**, 309–29.

[14] Kassapoglou, C. and Kaminski, M. (2011) Modeling damage and load redistribution in composites under tension-tension fatigue loading. *Composites Part A*, **42**, 1783–1792.

[15] Kassapoglou, C. (1990) Determination of interlaminar stresses in composite laminates under combined loads. *J. Reinf. Plast. Compos.*, **9**, 33–59.

[16] Kassapoglou, C. (2013) *Design and Analysis of Composite Structures*, 2nd edn, Chapter 3.2.1, John Wiley & Sons, Inc., Hoboken, NJ.

[17] Kassapoglou, C. (2013) *Design and Analysis of Composite Structures*, 2nd edn, Chapter 9.2.2, John Wiley & Sons, Inc, Hoboken, NJ.

[18] Berthelot, J.-M., Leblond, P., El Mahi, A. and Le Corre, J.-F. (1996) Transverse cracking of crossply laminates: part 1. Analysis. *Composites Part A*, **27**, 989–1001.

4

Delaminations

4.1 Introduction

Any separation between two adjacent plies in a lay-up is called a *delamination* (Figure 4.1). A delamination occurs any time an out-of-plane load causes local inter-laminar stresses that exceed the strength of the thin matrix layer between plies.

The out-of-plane loads can be mechanical or hygrothermal. Mismatch in Poisson's ratios, coefficients of mutual influence (ratio of in-plane axial to out-of-plane shear strains) or swelling coefficients can cause local out-of-plane stresses. In addition, delaminations can occur as a result of other local failures such as matrix cracks. For example, matrix cracks in adjacent plies of different orientations may cause a delamination at the ply interface. This is shown schematically in Figure 4.2, where a 90° ply is shown above a 0° ply. Even though the static loading does not cause local damage which will lead to a delamination, it is likely that fatigue loading will.

For some geometries such as the one shown in Figure 4.3, matrix cracks can evolve into delaminations under fatigue loads. Even low cyclic loads eventually crack the matrix in the resin pocket. These matrix cracks can coalesce into delaminations between the flange and skin of Figure 4.3.

Some of the most common situations where local out-of-plane stresses arise that may lead to delamination are shown in Figure 4.4. The bold (red) lines in Figure 4.4 indicate critical locations where delaminations may occur. Case (a) corresponds to an external plydrop or a flange/skin intersection of a stiffened panel. In general, a delamination will appear at the interface between the terminated and continuous plies. However, inter-laminar stresses will also be present near the edge of the dropped plies, and they too may cause delamination. This is covered in Section 4.3.5. Case (b) is the case of a straight free edge in a composite coupon covered in Section 4.3.4. Case (c) is the case of a sandwich ramp-down discussed briefly in [1]. In Case (d), resin pockets are created at internally terminated plies. The load transfer to adjacent continuous plies causes inter-laminar normal and shear stresses that can cause delamination. Case (e) is another free edge case, where, in addition to the straight edges of the coupon, there is the circular free edge of the hole where delaminations may occur.

Delaminations can also be created during manufacturing. Two relatively common situations are shown in Figure 4.5. In the first case on the left of Figure 4.5, contour

Modeling the Effect of Damage in Composite Structures: Simplified Approaches, First Edition. Christos Kassapoglou.
© 2015 John Wiley & Sons, Ltd. Published 2015 by John Wiley & Sons, Ltd.

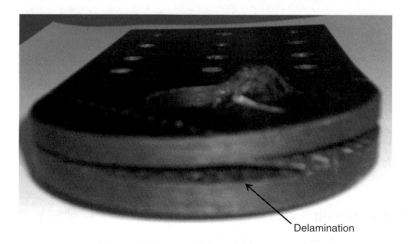

Delamination

Figure 4.1 Delamination in a composite lug after transverse loading (*See insert for colour representation of this figure.*)

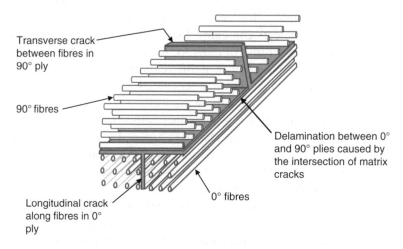

Transverse crack between fibres in 90° ply

90° fibres

Delamination between 0° and 90° plies caused by the intersection of matrix cracks

Longitudinal crack along fibres in 0° ply

0° fibres

Figure 4.2   Intersecting matrix cracks causing a delamination at the ply interface

mismatch between the skin and rib of a wing leads to gaps between the two during assembly. As fasteners are tightened to remove the gaps, delaminations may appear emanating from the fastener holes.

In the second case, on the right of Figure 4.5, a laminate is laid up in a concave (female) tool. Plies with fibres aligned with the circumferential direction of the tool do not easily conform to the mould shape especially if the corner radius is relatively small (<1 cm). 'Bridging' occurs where the ply stretches or jumps from one wall of the mould to the other without closely following the contour. Gaps are created, which, even if filled with resin during cure, eventually become delaminations as the resin cracks under repeated loading.

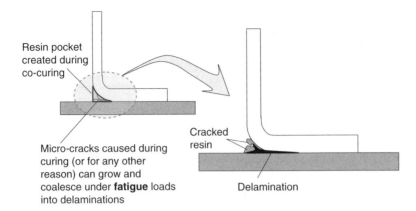

Figure 4.3   Cracks in resin pockets evolving to delamination (*See insert for colour representation of this figure.*)

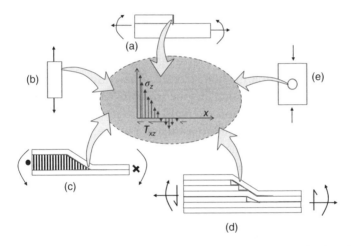

Figure 4.4   Structural details where local inter-laminar stresses may lead to delaminations (*See insert for colour representation of this figure.*)

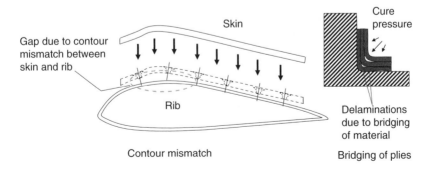

Figure 4.5   Creation of delaminations during manufacturing

Finally, the last of the main causes of delamination creation is impact during manufacturing, service or maintenance. Tool drops, foot traffic, runway debris, luggage drops and hail damage are some of the causes of impact damage, which includes, among others, delaminations below the impact location. Impact damage is discussed in more detail in Chapter 5.

4.2 Relation to Inspection Methods and Criteria

As can be seen from the preceding discussion, delaminations are unavoidable in a composite structure. It is, therefore, important to design structures that can continue successful operation with delaminations present. Within this context, the size of the delamination becomes very important. Very small delaminations may not even be detectable by the non-destructive inspection method selected. Then, the structure must be able to meet ultimate load with such delaminations present. In addition, under cyclic loading these delaminations should not grow to a size that could affect the performance by reducing load capability below limit load.

In broad terms, one can define a threshold and a critical delamination size. The threshold size is the smallest possible size that can be detected reliably (95–99% of the time) with the inspection method selected. The critical delamination size is the largest size that can be present such that the structure can meet the ultimate load and can last for at least two inspection intervals before it is detected and repaired. The two inspection intervals correspond to a situation where, during one inspection, the delamination is a little smaller than the threshold size and thus is not detected. Then, after one inspection interval, during the next inspection, for some reason the delamination is not detected even though it may be larger than the threshold size. Allowing inspection methods to miss a detectable delamination once adds conservatism to the structure. After one more inspection interval, it is expected that the delamination will be detected and will be repaired, or otherwise dispositioned (for example, 'use as is'). The size of the delamination at this stage cannot be greater than the critical delamination size. Safety factors can be imposed on these sizes depending on the situation and design practices.

In the discussion above, neither the inspection method nor the threshold and critical sizes were defined. The reason is that the sizes change with the inspection method. More elaborate and expensive methods can detect much smaller delaminations. A schematic with approximate ranges for the smallest detectable size for different inspection methods is shown in Figure 4.6.

Ideally, inspections would be carried out in the field with easy-to-use portable equipment. Unfortunately, the more accurate inspection methods such as X-rays are very expensive and not easily usable in the field. A compromise is usually reached between the smallest detectable delamination size and the cost and ease of use of equipment. Typically, hand-held ultrasonic techniques are used. This choice places the smallest detectable size in the range of a few millimetres. There are significant differences

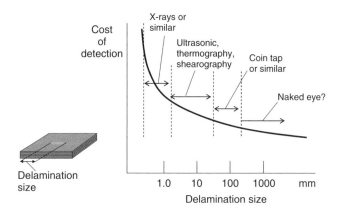

Figure 4.6 Approximate relation between the inspection method, detectable size and cost

in the smallest detectable delamination size in flat laminates as opposed to curved laminates. Delaminations as small as 4–6 mm in diameter can be detected reliably in flat parts. In curved parts such as 90° corners, the corresponding size is almost double and it requires special equipment that guides the detection probe to follow accurately the contour of the part.

In addition to ultrasonic inspection, visual inspection is used. When it comes to delaminations, visual inspection can only detect large delaminations and only after one of the delaminating sub-laminates has separated sufficiently from the laminate to appear as a 'bump' of some kind. In between ultrasonic and visual inspection, the 'coin tap' method is often used on specific parts. A tap hammer, which, for rough evaluation, used to be replaced by a coin, and hence the name of the method, is used to lightly tap on the structure. Changes in the sound heard indicate the presence of delaminations. Experienced inspectors can detect delaminations as small as 2 cm in diameter. However, the reliability and accuracy of this method is not high.

Delaminations are a good example of a situation where the design and analysis methods must interact with the inspection and repair techniques selected. The analysis methods must be able to predict whether a delamination with threshold size will grow under a given load. If a delamination grows, the analysis method must be able to determine whether this growth is stable or unstable and, on the basis of that, determine the critical size. For that, growth under cyclic loading must also be taken into account.

Briefly, the procedure in designing a composite structure with delaminations is as follows:

1. Based on previous experience and cost considerations, an inspection method is selected, for example, ultrasonic inspection.
2. The threshold delamination size is determined for different configurations (flat versus curved, with or without plydrops, etc.). The size may not be a single size for all types of parts. For ultrasonic inspection, typical size is in the range of 5–8 mm.

3. Develop analysis methods that determine whether a delamination with size equal to the threshold size will grow under the given static loads.
4. Establish whether growth is stable or unstable. If unstable, ensure that the structure can meet the ultimate load requirements with a threshold size delamination present.
5. Use tests and analysis if available to quantify delamination growth under cyclic loads (see Figure 4.7).
6. Combine the results of steps 4 and 5 to determine the critical delamination size.
7. Apply any additional safety factors to the critical delamination size. In extreme cases, one can set the threshold size equal to the critical size. This means it must be demonstrated that there is no growth of a threshold size delamination for the life of the structure.
8. Demonstrate that a delamination with size equal to the threshold size will not grow to a size greater than the critical size after two inspection intervals. Establish inspection intervals so that this requirement can be met.

If during inspection a delamination is found, it must be dispositioned. There are three possible dispositions: (i) Use 'as is'. In this case, the delamination is smaller than the threshold size and there is reliable information that the delamination will not grow at all or will grow slowly so that it will not exceed the critical size after two inspection intervals. (ii) Cosmetic repair. This is similar to the previous case but, for reasons not directly associated with safety, a repair that does not necessarily restore the pristine strength and stiffness capabilities is used. One example is a situation where a delamination may be at a location and of a size that do not affect performance but are visible to passengers and is repaired for psychological reasons. (iii) Structural repair. If the delamination is greater than the threshold size, it must be repaired in a way that the strength and stiffness of the pristine structure are restored. This is not easy because simply injecting resin and/or adhesive between the delaminating layers does not guarantee a bond between them that is at least as good as the original bond before the delamination was created.

Delamination growth under cyclic loading will be addressed, to some extent, in Chapter 6, see, for example Figures 6.1 and 6.2. For now, it should be pointed out that,

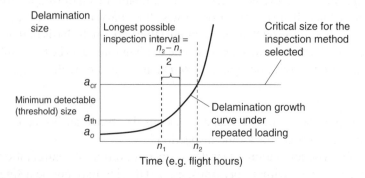

Figure 4.7   Schematic of delamination growth under cyclic loading

typically, growth, when it happens under cyclic loading, is quite rapid. This means that establishing inspection intervals on the basis of the approach just described and test data with trends similar to those in Figure 4.7 may result in very short inspection intervals. In addition, the experimental scatter associated with such tests and tests needed to determine model parameters, such as critical energy release rate in mode II for example, is large, requiring a large database to establish reliable inspection intervals. Finally, the analysis methods themselves (see subsequent sections) are still under development and, in many cases, not advanced enough to give reliable predictions for all but the simplest of structural configurations. For all these reasons, the preferred approach is a no-growth approach where it is demonstrated that a delamination of threshold size does not grow for the life of the structure appropriately adjusted to include safety factors and to account for experimental scatter.

4.3   Modelling Different Structural Details in the Presence of Delaminations

Analysis methods recognise that a delamination divides a laminate in two sub-laminates and attempt to determine the structural response of each sub-laminate as well as the interaction between the two at their interface where the delamination may grow. For more accurate modelling, the thin resin layer between the two sub-laminates should be modelled. Under in-plane loading, local stresses and strains in the two sub-laminates are used to determine whether one or both of the sub-laminates will buckle and whether the delamination will grow. Under out-of-plane loading, fracture mechanics approaches are mostly used to quantify delamination growth. Examples of these will be given in subsequent sections.

4.3.1   Buckling of a Through-Width Delaminating Layer

One of the simplest problems to give some insight into how a delamination may affect structural performance is the one-dimensional problem of a through-width delamination in a flat rectangular composite laminate. The situation is shown in Figure 4.8.

In this case, the delamination extends from one edge of the laminate to the other in the direction perpendicular to the applied load. The laminate is loaded in compression by a load per unit width N_x and the delamination divides the laminate in a relatively thin delaminating layer and a thick substrate. This is a 'thin-film' delaminating layer which will buckle early before the substrate buckles. Cases where the two delaminating layers may be of comparable thickness have been studied by Kardomateas and Schmueser [2].

It is assumed that the delamination is long in the y-direction so there is no dependence on y; therefore, $\partial/\partial y = 0$. The total applied load per unit width N_x is then split into N_{xd} (the load on the delaminating layer) and N_{xs} (the load on the substrate),

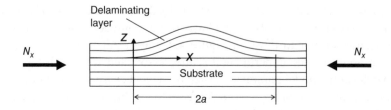

Figure 4.8   Through-width delamination

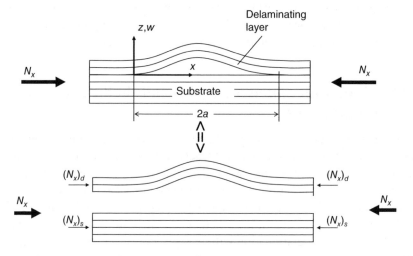

Figure 4.9   Sub-laminates created by a through-width delamination

see Figure 4.9, such that

$$N_x = (N_x)_s + (N_x)_d \tag{4.1}$$

With reference to the coordinate system at the top of Figure 4.9, and assuming that the edges of the delaminating layer are simply supported, the out-of-plane displacement w of the delaminating layer can be written as

$$w = w_o \sin \frac{\pi x}{2a} \tag{4.2}$$

with $2a$ the delamination length and w_o the unknown amplitude of w.

Note that, for typical delaminations in a composite laminate, the boundary condition is closer to clamped. The use of simply-supported boundary conditions is conservative.

In general, the delaminating layer will be unsymmetric, so its B matrix will be non-zero. An approximate way to account for unsymmetric lay-up in the delaminating layer is to use the 'reduced' D matrix for that laminate denoted with an overbar:

$$(\overline{D_{ij}})_d = (D_{ij})_d - (B_{ij})_d (A_{ij})_d^{-1} (B_{ij})_d \tag{4.3}$$

where the subscripts ij ($i, j = 1, 2, 6$) denote entries of the corresponding matrices. The inverse of the reduced D matrix relates the laminate curvatures to bending moments for a general laminate where the B matrix is non-zero [3].

To a first order, the delamination in Figure 2.9 does not affect the laminate performance until it buckles. This is valid for symmetric sub-laminates, but only approximately so for unsymmetric sub-laminates. In the latter case, small compressive loads lead to bending and/or twisting of the delaminating layer before the delaminating layer buckles.

Therefore, the buckling load of the delaminating layer can be used as an approximate design load, which, in conjunction with the simply-supported boundary condition of the delamination edges, makes for a conservative design. The buckling load of the delaminating sub-laminate is determined using energy minimisation. The energy stored minus the work done is given by

$$\Pi_p = \frac{1}{2} \int_0^{2a} \int_0^b \overline{(D_{11})}_d \left(\frac{\partial^2 w}{\partial x^2} \right)^2 dx dy + \frac{1}{2} \int_0^{2a} \int_0^b N_{xd} \left(\frac{\partial w}{\partial x} \right)^2 dx dy \tag{4.4}$$

where N_{xd} is the applied load on the delaminating layer, see also Figure 4.9.

As a result of the assumption that $\partial/\partial y = 0$, the problem has become a one-dimensional problem. Using Equation 4.2 to substitute in Equation 4.4 permits evaluation of the two integrals:

$$\int_0^{2a} \left(\frac{\partial^2 w}{\partial x^2} \right)^2 dx = \frac{w_o^2 \pi^4}{16a^3} \tag{4.5}$$

$$\int_0^{2a} \left(\frac{\partial w}{\partial x} \right)^2 dx = \frac{w_o^2 \pi^2}{4a} \tag{4.6}$$

which, in turn, lead to the following energy expression:

$$\Pi_p = \frac{1}{2} \overline{(D_{11})}_d \frac{w_o^2 \pi^4 b}{16a^3} + \frac{1}{2} (-N_{xbuck})_d \frac{w_o^2 \pi^2 b}{4a} \tag{4.7}$$

where $(-N_{xbuck})_d$ was substituted for N_{xd}. This is done for convenience. A positive value of $(N_{xbuck})_d$ implies compression. To minimise the energy, the right-hand side of Equation 4.7 is differentiated with respect to w_o and the result is set equal to zero:

$$\frac{\partial \Pi_p}{\partial w_o} = 0 \Rightarrow \frac{2w_o \pi^2 b}{4a} \left[\frac{\pi^2}{4a^2} \overline{(D_{11})}_d - (N_{xbuck})_d \right] = 0 \tag{4.8}$$

Solving Equation 4.8 for the buckling load of the delaminating layer, and assuming $w_o \neq 0$ (non-trivial solution), gives

$$(N_{xbuck})_d = \frac{\pi^2}{4a^2} \overline{(D_{11})}_d \tag{4.9}$$

Equation 4.9 has many similarities with the buckling load per unit width of a simply-supported beam:

$$N_{xcrit} = \frac{\pi^2 EI}{bL^2} \tag{4.10}$$

If the length of the delaminating layer $2a$ is set equal to the length of the beam L, and the bending stiffness per unit width $(\overline{D}_{11})_d$ is set equal to EI/b where b is the beam width perpendicular to the page in Figure 4.9, the two expressions are identical. Thus, the result of Equation 4.9 could have been anticipated from the buckling theory of beams.

It is important to note that Equation 4.9 only gives the 'local' load that would cause buckling of the delaminating layer. This load must be related to the load N_x applied to the entire laminate.

Looking at the bottom part of Figure 4.9, strain compatibility gives

$$\varepsilon_x = (\varepsilon_x)_d = (\varepsilon_x)_s \tag{4.11}$$

Equation 4.11 states that, at the edge of the delamination, the axial strain in the entire laminate is equal to the corresponding strains in the delaminating layer and the substrate. These strains can be related to the applied compressive loads via

$$(\varepsilon_x)_d = (\alpha_{11})_d (N_x)_d$$
$$(\varepsilon_x)_s = (\alpha_{11})_s (N_x)_s \tag{4.12}$$

where α_{11} is the 11 entry of the inverse of the entire ABD matrix. It is equal to a_{11} the inverse of the A matrix only if the B matrix is zero (symmetric delaminating layer and/or substrate).

Combining Equations 4.11 and 4.12 with Equation 4.1 in order to relate $(N_x)_d$ and $(N_x)_s$ to N_x gives

$$(N_x)_s = \frac{(\alpha_{11})_d}{(\alpha_{11})_s}(N_x)_d \tag{4.13}$$

and

$$(N_x)_d = \frac{(\alpha_{11})_s}{(\alpha_{11})_s + (\alpha_{11})_d} N_x \tag{4.14}$$

Finally, combining Equations 4.14 and 4.9, and recognising that at buckling the magnitude of $(N_x)_d$ equals the buckling load (N_{xbuck}) of the delaminating layer, gives the applied load N_x that would cause the delaminating layer to buckle:

$$N_{xbuck} = \frac{(\alpha_{11})_s + (\alpha_{11})_d}{(\alpha_{11})_s} \frac{\pi^2}{4a^2}(\overline{D}_{11})_d \tag{4.15}$$

An example is used to see how Equation 4.15 relates to other failure modes in a composite laminate. A quasi-isotropic [45/−45/0/90]s laminate is used with basic material properties, as shown in Table 4.1.

Table 4.1 Material properties

Property	Value
E_x(GPa)	131.0
E_y(GPa)	11.37
G_{xy}(GPa)	5.17
V_{xy}	0.29
t_{ply}(mm)	0.305

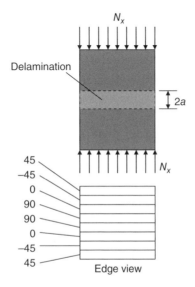

Figure 4.10 Compression-loaded quasi-isotropic laminate with through-width delamination

The laminate is loaded under compression and has a through-width delamination, as shown in Figure 4.10. The location of the delamination through the thickness is unknown.

If the location of the delamination through the laminate thickness is unknown, all possible locations must be examined. As this is a symmetric laminate, it suffices to check the cases with a delamination at each of the first four ply interfaces. This means determining the load N_x that would cause buckling of the delamination and comparing that to the most critical failure load if the delamination were not present. For simplicity, it is assumed that the laminate dimensions are relatively small and that the overall buckling load for the laminate is higher than the load corresponding to material failure. The material failure is, conservatively, taken to correspond to a strain cut-off of 4500 microstrain [4].

Equation 4.15 is used to determine the value of N_x at which the delaminating layer will buckle as a function of delamination length and location through the thickness.

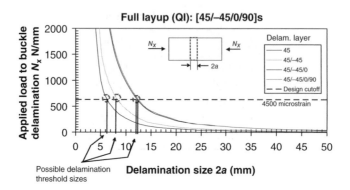

Figure 4.11   Buckling load of delaminating layer compared to material failure

The results are shown in Figure 4.11. Note that, for the case of a delamination at the mid-plane, the delaminating layer and the substrate are identical so both will buckle at the same time.

For each through-the-thickness location of the delamination, a different curve is shown in Figure 4.11. Obviously, as the delamination size increases, the buckling load for the delaminating layer decreases. As the delamination location moves deeper into the laminate and the delaminating layer becomes thicker and, as a result, stiffer, the buckling load for the delaminating layer increases. Note that the curves corresponding to a delamination at the third ply interface (delaminating layer 45/−45/0) and at the mid-plane (delaminating layer is half the laminate) are almost identical. The reason is that the difference between a delamination in the third and fourth ply interface is a 90° ply which has very little stiffness and its thickness adds little to the thickness of the three-ply delaminating layer. Therefore, the increase in stiffness from a delaminating layer of three to four plies, in this particular case, is negligible and the two buckling curves are very close to each other. It should be noted here that it is very unlikely that a delamination would occur at the mid-plane of this laminate as the two plies next to the mid-plane are identical and there is no pure resin layer between them, or its thickness is much smaller and is not as well defined as between plies of different orientations. It is, therefore, unlikely that a delamination would occur there unless some foreign object (for example, a peel ply) was left between the two 90 plies during manufacturing.

The material strength curve is a horizontal line in Figure 4.11, independent of the delamination size. The points of intersection of the material strength curve with the buckling curves define delamination sizes that are important in design. The inspection method selected must be able to reliably detect these sizes. The smallest size is a little larger than 6 mm and corresponds to a delamination at the first ply interface. As mentioned in Section 4.2, this happens to be in the range of reliably detectable delamination sizes when ultrasonic methods are used. If this size were significantly smaller, on the order of 1–2 mm, then ultrasonic methods might not be able to detect it reliably. In such a case, one would have to show that the laminate with a delamination present at the first ply interface of size equal to the smallest reliably detectable (4–6 mm) would meet the ultimate load. If that were not possible, the inspection method or the design would have to change.

In general, safety factors may be applied to the delamination sizes of Figure 4.11. The laminate may be designed assuming delaminations greater than these sizes. The safety factors used, if any, depend on the criticality of the structure and the type of application.

As already mentioned, the results obtained in this example are approximate. The delaminating layer is unsymmetric for most of the delamination locations, which is not explicitly accounted for by the analysis other than the use of the reduced bending stiffness matrix of Equation 4.3. Furthermore, the substrate is also unsymmetric, and for thick delaminating layers it may no longer stay straight before buckling of the delaminating layer, thus violating a basic assumption of the previous derivation. Also, the loaded edges of the delamination are assumed to be simply supported, which is conservative. A more detailed treatment of this problem can be found in [5].

4.3.2   Buckling of an Elliptical Delaminating Layer

The discussion in the previous section focussed on a simple example not often encountered in practice. Completely embedded delaminations are more common. In particular, delaminations created during service due to impact or fatigue loading are enclosed in the structure and tend to have an elliptical shape. Even if the shape is not elliptical, conservatively, an elliptical delamination circumscribing the actual delamination found in the structure can be used for modelling purposes. This is shown in Figure 4.12 for the case of a laminate under compression.

Again, as in the case of through-width delamination, the delaminating layer is assumed to have lower stiffness than the substrate, so buckling of the delaminating layer occurs first. The discussion here follows the approach by Chai and Babcock [6] and Kassapoglou and Hammer [7].

The elliptical delamination has major axis $2a$ (parallel to the x-axis) and minor axis $2b$. The exact boundary conditions at the delamination edge are not known. Test

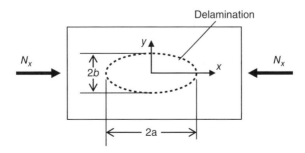

Figure 4.12   Elliptical delamination embedded in a laminate under compression

results have shown that the edge of the delaminating layer is closer to clamped than simply supported. For a conservative analysis, the boundary of the delamination can be assumed as simply supported.

The buckling load of the delaminating layer is obtained by a Rayleigh–Ritz approach. The out-of-plane displacement w of the delaminating layer is assumed as a power series, and the energy in the delaminating layer is minimised with respect to unknown coefficients in w.

For a simply-supported delamination boundary, the following expression can be used:

$$w = \left(1 - \frac{x^2}{a^2} - \frac{y^2}{b^2}\right)(w_{00} + w_{10}x^2 + w_{01}y^2 + w_{11}x^4 + \cdots) \tag{4.16}$$

The first term in parenthesis on the right-hand side is the equation of an ellipse with semi-major and semi-minor axes a and b, respectively. Thus, it is zero on the boundary of the delamination satisfying the simply-supported boundary condition.

For clamped delamination boundaries, the corresponding expression is

$$w = \left(1 - \frac{x^2}{a^2} - \frac{y^2}{b^2}\right)^2(w_{00} + w_{02}x^2 + w_{20}y^2 + w_{40}x^4 + \cdots) \tag{4.17}$$

In this case, the first term in parentheses on the right-hand side is squared, guaranteeing that both the deflection and the slopes $\partial w/\partial x$ and $\partial w/\partial y$ are zero on the boundary, consistent with clamped boundary conditions. The unknowns w_{00}, $w_{10,}$ w_{01}, and so on, are determined by energy minimisation.

It these simpler forms, there are no odd powers of x and y in Equations 4.16 and 4.17. This means that the buckled shape of the delaminating layer is assumed symmetric with respect to the x- and y-axis. This would be a valid assumption if the D_{16} and D_{26} terms of the delaminating layer were negligible. The number of terms in the series in the second parenthesis on the right-hand side of Equations 4.16 and 4.17 depends on the required level of accuracy. Typically, the first five terms give satisfactory accuracy, within at worst 10% of the converged solution when more terms are used and when the delamination aspect ratio is in the range $0.5 < a/b < 2$. The energy expression to be minimised is [8]

$$\Pi_c = \frac{1}{2}\iint\left\{\begin{array}{l}D_{11}\left(\frac{\partial^2 w}{\partial x^2}\right)^2 + 2D_{12}\frac{\partial^2 w}{\partial x^2}\frac{\partial^2 w}{\partial y^2} + D_{22}\left(\frac{\partial^2 w}{\partial y^2}\right)^2 + 4D_{66}\left(\frac{\partial^2 w}{\partial x\partial y}\right)^2 \\ + 4D_{16}\frac{\partial^2 w}{\partial x^2}\frac{\partial^2 w}{\partial x\partial y} + 4D_{26}\frac{\partial^2 w}{\partial y^2}\frac{\partial^2 w}{\partial x\partial y}\end{array}\right\}dxdy$$

$$+ \frac{1}{2}\iint N_x\left(\frac{\partial w}{\partial x}\right)^2 dxdy + \frac{1}{2}\iint N_y\left(\frac{\partial w}{\partial y}\right)^2 dxdy + \iint N_{xy}\frac{\partial w}{\partial x}\frac{\partial w}{\partial y}dxdy \tag{4.18}$$

Note that in Equation 4.18 the general in-plane loading where N_x, N_y and N_{xy} are all non-zero is represented. Also, laminates with non-zero D_{16} and D_{26} can be modelled.

Using Equation 4.16 or 4.17, depending on the boundary conditions selected, to substitute in Equation 4.18, performing the integrations, differentiating with respect to the unknowns w_{00}, w_{10}, and so on, and setting the derivatives to zero leads to a generalised eigen-value problem of the form

$$[C]\{x\} = \lambda[F]\{x\} \qquad (4.19)$$

where [C] is a matrix involving the integrals in Equation 4.18, which include the D matrix, $\{x\}$ is the eigenvector of unknown coefficients

$$\{x\}^T = \{w_{00} \; w_{02} \; w_{20} \; w_{40} \; w_{22} \; ...\}$$

and [F] is a matrix involving the problem geometry and the relative magnitudes of N_x, N_y and N_{xy}.

Note that, because of the fact that the integrations have as limits the boundaries of the ellipse, it is expedient to perform the integrations numerically using Gaussian quadrature. Following Cairns [9], a 21×21 point Gaussian quadrature is used here, which would integrate exactly any polynomial up to order 10. This means that, if orders higher than 10 are used in Equation 4.17, more Gaussian integration points should be used.

The lowest eigenvalue λ of Equation 4.19 gives the buckling load. For the general case, where N_y and N_{xy} are non-zero, all entries of the load vector $\{N_x \; N_y \; N_{xy}\}^T$ are multiplied by λ.

As an example, consider the same laminate as in the previous section but, this time, instead of a through-width delamination, an embedded elliptical delamination is present. The aspect ratio of the delamination is $a/b = 1.25$. The delamination is placed successively in each ply interface, and the load N_x applied to the entire laminate to cause buckling of the delamination is determined using the approach just described. For this example, the unknowns in Equation 4.17 were w_{00}, w_{20}, w_{02}, w_{11} and w_{22}. The results are shown in Figure 4.13.

The results in this figure can be compared with those in Figure 4.11. It should be pointed out that the cut-off load is slightly different in the two figures even though the cut-off strain is the same, namely 4500 microstrain. The reason is that the results in Figure 4.11 are for a one-dimensional case where

$$N_x = A_{11}\varepsilon_x$$

while in Figure 4.13 they correspond to a two-dimensional case where the Poisson's ratio effect is present:

$$N_x = \left(A_{11} - \frac{A_{12}^{\,2}}{A_{22}}\right)\varepsilon_x$$

As should be expected, the delamination threshold sizes for the elliptical delamination in Figure 4.13 are significantly larger than those for a through-width delamination in Figure 4.11. There are two reasons for this: First, the elliptical delamination was

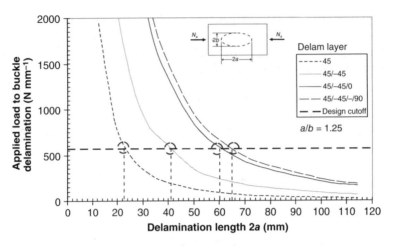

Figure 4.13 Load to cause buckling of delaminating layer in [45/−45/0/90]s laminate (*See insert for colour representation of this figure.*)

assumed to be clamped, while the through-width delamination was assumed simply supported. Second, the two-dimensional elliptical delamination is constrained all around its boundary and will have a higher buckling load than a delamination unconstrained along the two unloaded edges.

A comparison of analytical predictions to test results follows. In order to stay within the assumptions of thin-film theory, that is, the buckling load of the delaminating layer is much lower than the buckling load of the substrate, sandwich specimens were fabricated with 19 mm (0.75 in.) thick honeycomb and various facesheet lay-ups, shown in Table 4.2. The specimens were 533.4 mm long by 152.4 mm wide. The core at the two ends of the specimens was removed and replaced by solid aluminium blocks (152.4 mm on either side) to facilitate gripping and load introduction without crushing the core. Two different delamination sizes were used: 25.4 mm × 19.05 mm and 57.15 mm × 50.18 mm, with the longest dimension aligned with the load direction and the 0 direction of the lay-up. The delaminations were located between the first ply next to the core and its neighbour. The analytical predictions were obtained using Equation 4.17 for clamped delamination edges, with the following unknowns in the w expression: w_{00}, w_{20}, w_{11}, w_{02} and w_{22}. A strain compatibility condition was used at the delamination edge to relate the buckling load of the delaminating layer to the total applied load on the sandwich.

The test results are compared to the analytical predictions in Table 4.3. Interestingly, the predictions for two of the larger sized delaminations are lower than the test results. For all remaining cases, the predictions are higher than the test results, which is expected since the model used here uses only five unknowns and should be stiffer than the test specimens. Irrespective of whether they are conservative or unconservative, the predictions are within 12% of the test results. This suggests that the buckling analysis presented in this section can be used for preliminary design of flat laminates with delaminations under in-plane loads.

Table 4.2 Facesheets lay-ups in sandwich specimens

Specimen designation	Facesheet lay-up
2	5HS(45)/T0/W(0)/T0/5HS(45)
5	W(45)/T0/5HS(45)/T0/W(45)
6	W(45)/T0/W(45)/T0/W(45)
21	W(45)/W(45)/W(45)/W(45)

The zero direction is aligned with the load. 5HS: five-harness satin 0.3683 mm thick; T: unidirectional tape 0.1397 mm thick; W: plain weave fabric 0.2286 mm thick.

Table 4.3 Buckling of delaminating layer compared to test results

Panel	Test (N mm^{-1})	Prediction (N mm^{-1})	Percent difference
25.4 mm × 19.05 mm delamination			
2	379.5	421.7	10.0
5	414.0	429.6	3.6
6	422.0	433.9	2.7
57.15 mm × 50.8 mm delamination			
2	305.2	294.2	−3.7
5	316.6	303.0	−4.5
6	379.7	402.8	5.7
21	269.0	305.6	12.0

4.3.3 Application – Buckling of an Elliptical Delamination under Combined Loads

Consider the case of a [45/−45/0/90/A] laminate, shown in Figure 4.14, where A is a stiff sub-laminate of any lay-up as long as the total laminate is symmetric. An elliptical delamination is located at the fourth ply interface, below the 90° ply. The delamination boundary is assumed to be clamped. The length $2a$ and width $2b$ are set to 25 and 20 mm, respectively. The longer dimension is parallel to the load N_x. The coordinate system is the same as in Figure 4.12.

The analysis follows the procedure outlined by Equations 4.17–4.19. However, since D_{16} and D_{26} of the reduced D matrix from Equation 4.3 are appreciable, Equation 4.17 now includes the odd powers of x and y:

$$w = \left(1 - \frac{x^2}{a^2} - \frac{y^2}{b^2}\right)^2 (w_{00} + w_{10}x + w_{01}y + w_{20}x^2 + w_{02}y^2) \qquad (4.17a)$$

This permits unsymmetric buckling modes which may manifest themselves due to the bending–twisting coupling present.

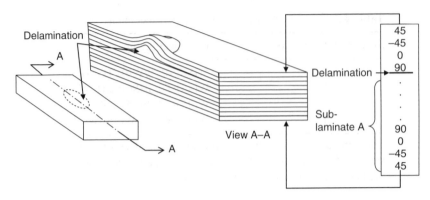

Figure 4.14 Laminate with embedded elliptical delamination (*See insert for colour representation of this figure.*)

For typical carbon epoxy material, the reduced D matrix of the delaminating sub-laminate [45/−45/0/90] is

$$D_{11} = 2505\,\text{N}\,\text{mm}$$
$$D_{12} = 1891\,\text{N}\,\text{mm}$$
$$D_{16} = 550\,\text{N}\,\text{mm}$$
$$D_{22} = 5142\,\text{N}\,\text{mm}$$
$$D_{26} = 768\,\text{N}\,\text{mm}$$
$$D_{66} = 966\,\text{N}\,\text{mm}$$

Solving the eigenvalue problem of Equation 4.19 leads to the results shown in Figure 4.15. The results correspond to biaxial compression combined with shear. Each curve in Figure 4.15 corresponds to a different applied shear load N_{xy} and defines the combination of N_x and N_y loads, which, for the given N_{xy} value, would cause the delaminating sub-laminate to buckle.

Load combinations inside a given curve, that is, closer to the origin, cause no buckling, while load combinations outside a given curve, further away from the origin, correspond to already buckled sub-laminate. It is interesting to note that small increases in N_{xy}, for example, from 0 to 2 N mm^{-1}, reduce the buckling load drastically, as seen by how close to the origin the curve for $N_{xy} = 2.0\,\text{N}\,\text{mm}^{-1}$ is compared to the curve for $N_{xy} = 0\,\text{N}\,\text{mm}^{-1}$. Also, for the particular configuration examined here, increasing N_{xy} decreases the N_x load-carrying ability much more than the ability to carry load N_y: Over the range of N_{xy} values examined, N_x decreases from 156 to 12 N mm^{-1}, while over the same N_{xy} range N_y decreases only from 250 to 230 N mm^{-1}.

As a final note, the buckling loads in Figure 4.15 do not correspond to the overall laminate loads. To translate them to laminate loads, one must use the specific sub-laminate A and compatibility relations analogous to Equation 4.12 which ensure

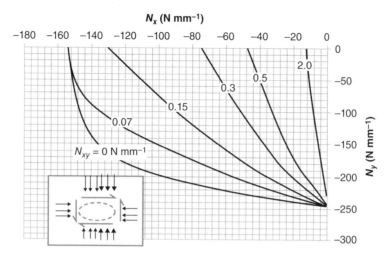

Figure 4.15 Buckling curves for an elliptical delaminating sub-laminate under combined in-plane loads

that the delaminating sub-laminate and the substrate have the same strains along the boundary of the delamination.

4.3.4 Onset of Delamination at a Straight Free Edge of a Composite Laminate

As mentioned at the beginning of this chapter, because of the mismatch in Poisson's ratios or coefficients of mutual influence, inter-laminar stresses develop at free edges of composite laminates. At sufficiently high loads, these stresses cause delaminations. The situation is shown schematically in Figure 4.16.

The best way to approach this problem is by determining the quantity that controls growth of the edge delamination in Figure 4.16. This is the strain energy release rate G. During a virtual crack growth in an elastic body, the crack driving force G, or strain energy release rate, equals the rate of decrease of strain energy U [10].

It is assumed that the delamination is past the nucleation stage so its area A is sufficient to quantify it. It is also assumed that the laminate in Figure 4.16 is linearly elastic and no body forces act. Typically, one differentiates between two cases: one where displacements are applied, and one where loads are applied; mixed displacements and loads will be discussed briefly later. If the tension load in Figure 4.16 is the result of an applied uniform displacement, then the strain energy U can be viewed as being dependent only on the delamination via its area A and the internal displacements u_i resulting from the applied displacement. In such a case, the incremental change in strain energy U can be written as

$$dU = -\left(\frac{\partial U}{\partial A}\right)_{u_i} dA + \left(\frac{\partial U}{\partial u_i}\right)_A du_i \qquad (4.20)$$

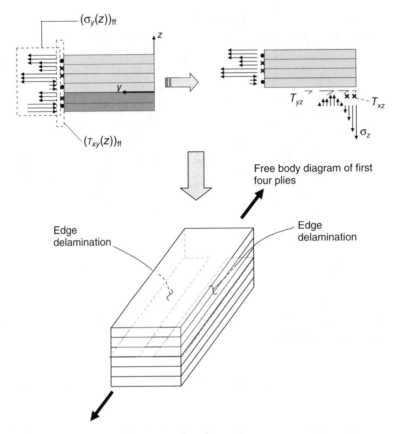

Figure 4.16   Delaminations developing at free edges of a laminate under tension

which is the total differential for the strain energy U. Note a minus sign is used in the first term in the left-hand side because, as the delamination area A increases, the strain energy U decreases. The subscripts 'u_i' and 'A' on the right-hand side of the equation denote that these quantities are kept constant. The strain energy release rate then is defined as the quantity multiplying dA on the right hand side of Equation 4.20:

$$G = -\left(\frac{\partial U}{\partial A}\right)_{u_i}$$
(4.21)

which is the rate of change of strain energy for an increase in delamination area with the displacements u_i held constant. Usually, the subscript u_i is dropped from the expression for G. Because of the fact that an increase in A reduces the strain energy of the laminate, the quantity $\partial U/\partial A$ is negative and the strain energy release rate G is always positive.

Instead of Equation 4.21, it is useful sometimes to express G in terms of the complementary energy U_p in the laminate. This would be a situation where, instead of an applied displacement, the load in Figure 4.16 is applied (controlled).

The complementary energy U_p in the laminate is defined as

$$U_p = \frac{1}{2} \int_{A_\sigma} \sigma_i u_i dA_\sigma \qquad (4.22)$$

where σ_i are the prescribed (controlled) stresses (or loads) on surface A_σ, and u_i are the corresponding displacements that result from these stresses. Note that body forces, which can be included in Equation 4.22 were neglected here.

For a linear system, the internal strain energy U equals the complementary energy U_p when the laminate is at equilibrium. However, while delamination propagation implies a decrease in the internal strain energy U, it implies an increase in the complementary energy U_p. This can be seen from Figure 4.17, where, for simplicity, force F is used instead of stress corresponding to displacement δ.

As the delamination grows incrementally from area A to $A + dA$, the load F follows a general curve as a function of δ, as shown in Figure 4.17. The load may be reduced from the original value of P, corresponding to point A, to a lower value corresponding to point A'. The actual relation $F(\delta)$ is, in general, unknown. The two extreme cases are the constant load case with constant P, in which case the change in displacement du is shown by segment AB, and the constant displacement case with constant u, as denoted by the excursion AC.

In general, the change in strain energy is the area of the shape OAA'O. As the shape of AA' is not known, in the limit, as du tends to zero, it can be approximated either by the area of the triangle OACO (constant displacement) or by the area of the triangle OABO (constant load P). Consider the latter case. The complementary energy before delamination growth is given by

$$U_p(u) = \int_0^{F(u)} \delta dF(\delta) = \text{Area(OAD)} = \frac{1}{2} Pu \qquad (4.23)$$

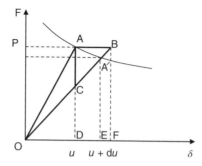

Figure 4.17 Force versus displacement relation during delamination growth

After delamination growth to $A + dA$, the displacement goes from u to $u + du$ and the complementary energy is given by

$$U_p(u + du) = \int_0^{F(u+du)} \delta dF(\delta) = \text{Area(OA'E)} \tag{4.24}$$

As du tends to zero, A' and B will coincide and Area(OA'E) $\rightarrow$ Area(OBF) and Equation 4.24 can be rewritten as

$$U_p(u + du) = \text{Area(OBF)} = \frac{1}{2}(u + du)P \tag{4.25}$$

Using Equations 4.24 and 4.25, the change in complementary energy as the delamination area increases from A to $A + dA$ can be found as

$$dU_p = U_p(u + du) - U_p(u) = \frac{1}{2}Pdu \tag{4.26}$$

which is valid under constant load P. Clearly, from Equation 4.26, as the delamination increases, the complementary energy also increases. Since, as already mentioned, the internal strain energy equals the complementary energy, Equation 4.21 can now be used along with the fact that the change in complementary energy is positive for a given increase in delamination area to obtain

$$G = \left(\frac{\partial U_p}{\partial A} \right)_P \tag{4.27}$$

Thus, the strain energy release rate under constant load can be determined using Equation 4.27.

As a note, in the general case where displacements and stresses may be applied over different parts of the boundary, Equations 4.21 and 4.27 can be generalised to

$$G = -\frac{\partial \Pi}{\partial A} \tag{4.28}$$

with

$$\Pi = U - \int_{A_s} \sigma_i u_i dA_s \tag{4.29}$$

The strain energy release rate G is a scalar quantity which combines the energy per unit area needed to grow a delamination by all three modes of loading, modes I, II and III, shown in Figure 4.18. For a given loading combination, that is, any mix of the three modes, there is a critical value of G, denoted by G_c, that will cause growth of a given delamination. The condition for growth is then

$$G = G_c \tag{4.30}$$

or, if each mode acts individually, Equation 4.30 takes the form

$$\begin{aligned} G &= G_{Ic} \\ G &= G_{IIc} \\ G &= G_{IIIc} \end{aligned} \tag{4.31a–c}$$

The values of G_c, G_{Ic}, G_{IIc} and G_{IIIc} are measured experimentally.

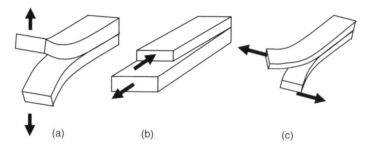

Figure 4.18   Three modes of loading a delamination: (a) opening, (b) shearing and (c) tearing

Returning to the problem at hand, the goal is to calculate the energy release rate of the laminate with edge delaminations in order to compare it to the critical energy release rate and determine the load at which there will be growth. This approach was pioneered by O'Brien [11]. The laminate is divided into different regions, as shown in Figure 4.19.

It is assumed that the laminate is symmetric, so if a delamination appears at a ply interface above the mid-plane, then it will also appear symmetrically at the corresponding ply interface below the mid-plane as well as at the opposite edge. Thus, in general, there are four edge delaminations created, as shown in Figure 4.19. Each delamination has length a. The delaminations divide the laminate into sub-laminates. The thickness of the ith sub-laminate is t_i. The width of the entire laminate is $2b$, and the thickness is h. Region A is the region containing the delaminations in one half of the laminate, and B is the corresponding in-tact region, as shown in Figure 4.19.

First, the equivalent stiffness of the laminate in the presence of delaminations is determined. Axial strain compatibility requires that the membrane stiffness EA (= Young's modulus times cross-sectional area) for any portion of the laminate is

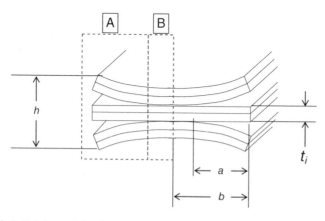

Figure 4.19   Subdivision of laminate under tension to regions based on the presence of delaminations

equal to the sum of the EA values of its individual constituents [12]. For the overall laminate, therefore

$$(EA)_{eq} = 2(EA)_A + 2(EA)_B \tag{4.32}$$

where the subscript 'eq' denotes the equivalent membrane stiffness for the entire laminate and the factors of 2 account for the two laminate halves (left and right) in Figure 4.19.

In an analogous manner

$$(EA)_A = \sum_{i=1}^{n} E_i t_i a \Rightarrow E_A a h = \sum_{i=1}^{n} E_i t_i a \tag{4.33}$$

where E_A is the overall axial stiffness of the delaminated portion of the laminate and

$$(EA)_B = E_{LAM}(b - a)h \tag{4.34}$$

where E_{LAM} is the stiffness of the laminate without delaminations.

At the same time

$$(EA)_{eq} = E_{eq} 2bh \tag{4.35}$$

Combining Equations 4.32–4.35 leads to an expression for the equivalent stiffness of the entire delaminated laminate, E_{eq}

$$E_{eq} 2bh = 2 \sum_{i=1}^{n} E_i t_i a + 2E_{LAM}(b - a)h \Rightarrow$$

$$E_{eq} = E_{LAM} + \left(\frac{\sum_{i=1}^{n} E_i t_i}{h} - E_{LAM} \right) \frac{a}{b} \tag{4.36}$$

Equation 4.36 forms the basis for the solution of the problem at hand. It relates the stiffness of the delaminated laminate to the stiffness of the laminate without delaminations and a 'correction factor' that is proportional to the delamination length a and the stiffnesses of the individual sub-laminates created by the delaminations. This equation can be combined with Equation 4.21 to obtain the energy release rate for this case.

In general, the strain energy is given by

$$U = \frac{1}{2} \iiint \sigma \varepsilon dV \tag{4.37}$$

where V is the laminate volume. For the particular case of a symmetric laminate under uniaxial tension, $\sigma = E\varepsilon$ and U becomes:

$$U = \frac{1}{2} \iiint E\varepsilon^2 dV = \frac{1}{2} \varepsilon^2 E_{eq} V \tag{4.38}$$

where the integration gives only a factor of V (the laminate volume) since the axial strain is constant throughout the laminate.

Combining Equations 4.28 and 4.21 gives

$$G = -\frac{1}{2}\varepsilon^2 2bhL\frac{dE_{eq}}{dA} \tag{4.39}$$

with L the laminate length and A the delamination area. For the case of a mid-plane delamination, the total delamination area is given by

$$A = 2aL \tag{4.40}$$

where the factor of 2 accounts for the left and right side of the laminate in Figure 4.18. Note that, if the delamination is not in the mid-plane, an additional factor of 2 must be included to account for all the delaminations above and below the mid-plane. Then

$$\frac{dE_{eq}}{dA} = \frac{1}{2L}\frac{dE_{eq}}{da} \tag{4.41}$$

Substituting in Equation 4.39 gives the final expression for the energy release rate:

$$G = \frac{1}{2}\varepsilon^2 h\left(E_{LAM} - \frac{\sum_{i=1}^{n}E_i t_i}{h}\right) \quad \text{(mid-plane delamination)} \tag{4.42a}$$

If the delamination is not at the mid-plane, Equation 4.42a has the form

$$G = \frac{1}{4}\varepsilon^2 h\left(E_{LAM} - \frac{\sum_{i=1}^{n}E_i t_i}{h}\right) \quad \text{(delamination away from mid-plane)} \tag{4.42b}$$

An important observation is in order. G in Equation 4.42a is independent of the delamination size. This means that it should hold for any delamination size including the limiting case of no delamination ($a = 0$). Then, Equation 4.42a can be used to determine the onset of delamination load. This is done by solving Equation 4.42a for the applied strain ε that would cause delamination onset and setting $G = G_c$ (the critical energy release rate)

$$\varepsilon_{crit} = \sqrt{\frac{2G_c}{h\left(E_{LAM} - \left(\sum_{i=1}^{n}E_i t_i/h\right)\right)}} \quad \text{(mid-plane delamination)} \tag{4.43a}$$

or

$$\varepsilon_{crit} = \sqrt{\frac{4G_c}{h\left(E_{LAM} - \left(\sum_{i=1}^{n}E_i t_i/h\right)\right)}} \quad \text{(delamination away from mid-plane)} \tag{4.43b}$$

The accuracy of Equation 4.43a in predicting onset of delamination was verified by O'Brien by comparing with test results. He carried out tests on a [±30/±30/90/90]s laminate to back-calculate the critical energy release rate for the onset of delamination at the mid-plane of this laminate. He then used this deduced value of G_c and Equation 4.43a to predict onset of mid-plane delamination in [45n/−45n/0n/90n]s

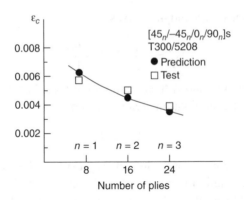

Figure 4.20   Onset of delamination strains for $[45n/-45n/0n/90n]s$ T300/5208 laminate

laminates as n is varied from 1 to 3. The results are shown in Figure 4.20. There is very good agreement between theoretical predictions and test results.

One issue that arises is: what exactly is the value to use for E_i in Equations 4.43a and 4.43b? In the simple, but limited, case where a sub-laminate is symmetric, its axial stiffness is given by [12]

$$E_i = \frac{1}{t_i(a_{11})_i}$$ (4.44)

where t_i is the sub-laminate thickness and $(a_{11})_i$ is the 11 entry of the inverse of the A matrix for the sub-laminate.

In a more general case where the sub-laminate is unsymmetric, the relation:

$$E_i = \frac{1}{t_i(\alpha_{11})_i}$$ (4.45)

can be used if the sub-laminate has $N_y = N_{xy} = M_x = M_y = M_{xy} = 0$. Here, α_{11} is the 11 entry of the inverse of the full ABD matrix of the laminate. This condition, where only N_x is non-zero for the sub-laminate, may not be general enough especially in cases where the asymmetry of the sub-laminates causes stretching–bending coupling and non-zero bending moments develop as a result of the applied tension load. In such cases, the actual boundary conditions of the structure play a role. For example, if the laminate in Figure 4.16 is constrained by the way the load is applied and by the surrounding structure to have $N_y = M_y = \gamma_{xy} = \kappa_x = \kappa_{xy} = 0$, the constitutive relationships for the sub-laminate have the form

$$\begin{Bmatrix} N_x \\ 0 \\ N_{xy} \\ M_x \\ 0 \\ M_{xy} \end{Bmatrix} = \begin{bmatrix} A_{11} & A_{12} & A_{16} & B_{11} & B_{12} & B_{16} \\ A_{12} & A_{22} & A_{26} & B_{12} & B_{22} & B_{26} \\ A_{16} & A_{26} & A_{66} & B_{16} & B_{26} & B_{66} \\ B_{11} & B_{12} & B_{16} & D_{11} & D_{12} & D_{16} \\ B_{12} & B_{22} & B_{26} & D_{12} & D_{22} & D_{26} \\ B_{16} & B_{26} & B_{66} & D_{16} & D_{26} & D_{66} \end{bmatrix} \begin{Bmatrix} \varepsilon_x \\ \varepsilon_y \\ 0 \\ 0 \\ \kappa_y \\ 0 \end{Bmatrix}$$ (4.46a–f)

where ε_x and ε_y are mid-plane strains of the sub-laminate.

Eliminating κ_y from Equations 4.46b and 4.46e allows the determination of ε_y in terms of ε_x:

$$\varepsilon_y = \frac{B_{11}B_{22} - A_{12}D_{22}}{A_{22}D_{22} - B_{22}{}^2}\varepsilon_x \tag{4.47}$$

Then, using one of the two Equation 4.46b or 4.46e, κ_y can be determined:

$$\kappa_y = \frac{A_{12}B_{22} - A_{22}B_{12}}{A_{22}D_{22} - B_{22}{}^2}\varepsilon_x \tag{4.48}$$

Now placing Equations 4.47 and 4.48 into Equation 4.46a gives N_x only as a function of ε_x:

$$N_x = \left[A_{11} + A_{12}\frac{B_{11}B_{22} - A_{12}D_{22}}{A_{22}D_{22} - B_{22}{}^2} + B_{12}\frac{A_{12}B_{22} - A_{22}B_{12}}{A_{22}D_{22} - B_{22}{}^2}\right]\varepsilon_x \tag{4.49}$$

Dividing both sides of Equation 4.40 by the sub-laminnate thickness t yields

$$\frac{N_x}{t} = \sigma_x = \frac{1}{t}\left[A_{11} + A_{12}\frac{B_{11}B_{22} - A_{12}D_{22}}{A_{22}D_{22} - B_{22}{}^2} + B_{12}\frac{A_{12}B_{22} - A_{22}B_{12}}{A_{22}D_{22} - B_{22}{}^2}\right]\varepsilon_x$$

from which the constant of proportionality relating stress to strain is the axial stiffness of the sub-laminate:

$$E = \frac{1}{t}\left[A_{11} + A_{12}\frac{B_{11}B_{22} - A_{12}D_{22}}{A_{22}D_{22} - B_{22}{}^2} + B_{12}\frac{A_{12}B_{22} - A_{22}B_{12}}{A_{22}D_{22} - B_{22}{}^2}\right] \tag{4.50}$$

It should be kept in mind that all quantities in Equation 4.50 refer to the ith sub-laminate. In a similar manner, other load–strain combinations can be analysed to determine the sub-laminate stiffness (see, for example [13]).

One final comment relates to the fact that G determined from Equation 4.42a is independent of the delamination size a. While this allows determination of the onset of delamination load, it contradicts the implicit assumption associated with the strain energy release rate that a delamination (or crack) is already present before load application and it grows by a certain amount when the load is applied. The situation where the delamination does not exist and suddenly jumps to a finite value is briefly addressed in Section 4.3.5. For now, it should be pointed out that in problems such as the delamination at a straight free edge, it is expected that, as the delamination becomes very small, some of the assumptions used in the derivation in this section break down. For example, Poisson's ratio effects and local out-of-plane effects were neglected in Equation 4.38. However, it is expected, and it is verified by test results, that the energy release rate goes from 0 to the value predicted by Equation 4.42a over a very short range of small delamination lengths, and therefore Equation 4.42a is valid only beyond these very small delamination sizes.

4.3.5   Delamination at a Flange–Stiffener Interface of a Composite Stiffened Panel

A very important structural detail that is prone to delamination is the flange–skin interface of composite stiffened panels. This is a complex three-dimensional problem with combined loading, especially when the skin is post-buckled. As a first step that may help in understanding delamination growth in such situations, a one-dimensional section cut of the flange and portion of the skin under a shear load V are shown in Figure 4.21. This is also the same as the external plydrop case, namely the pull-off specimen case, and, in addition, can be viewed as a variation of the single cantilever beam specimen proposed by Radcliffe and Reeder for sandwich facesheet disbond evaluation [14].

A delamination of length a is present between the flange and skin, as shown in Figure 4.21. The resin layer ahead of the delamination is shown by a bold black line. The flange and skin properties together are denoted by the subscript 1 and the skin properties by the subscript 2.

$$\text{Region::}\quad L_1 + a < x < L$$

The bending moment is given by

$$M = V(L - x) \tag{4.51}$$

Neglecting shear energy, the strain energy (due to bending only) can be written as

$$U_3 = \int_{L_1+a}^{L} \frac{M^2}{2(EI)_2} dx = \int_{L_1+a}^{L} \frac{V^2(L^2 - 2Lx + x^2)}{2(EI)_2} dx = \frac{V^2}{2(EI)_2} \left[L^2x - Lx^2 + \frac{x^3}{3} \right]_{L_1+a}^{L}$$

$$= \frac{V^2}{2(EI)_2} \left[\frac{L^3}{3} - L^2(L_1 + a) + L(L_1 + a)^2 - \frac{(L_1 + a)^3}{3} \right] \tag{4.52}$$

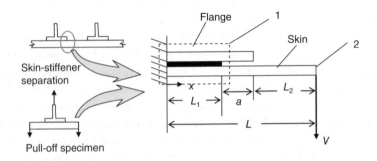

Figure 4.21   Delamination at a flange–skin interface

Similarly, in the region $L_1 < x < L_1 + a$, the strain energy is

$$U_2 = \int_{L_1}^{L_1+a} \frac{M^2}{2(EI)_2} dx = \int_{L_1}^{L_1+a} \frac{V^2(L^2 - 2Lx + x^2)}{2(EI)_2} dx = \frac{V^2}{2(EI)_2} \left[L^2 x - Lx^2 + \frac{x^3}{3} \right]_{L_1}^{L_1+a}$$

$$= \frac{V^2}{2(EI)_2} \left[L^2(L_1 + a) - L(L_1 + a)^2 + \frac{(L_1 + a)^3}{3} - L^2 L_1 + LL_1^2 - \frac{L_1^3}{3} \right] \quad (4.53)$$

And, in region $0 < x < L_1$

$$U_1 = \int_{0}^{L_1} \frac{M^2}{2(EI)_1} dx = \int_{0}^{L_1} \frac{V^2(L^2 - 2Lx + x^2)}{2(EI)_1} dx = \frac{V^2}{2(EI)_1} \left[L^2 x - Lx^2 + \frac{x^3}{3} \right]_{0}^{L_1}$$

$$= \frac{V^2}{2(EI)_1} \left[L^2 L_1 - LL_1^2 + \frac{L_1^3}{3} \right] \quad (4.54)$$

Equations 4.52–4.54 can be combined to get total strain energy U:

$$U = U_1 + U_2 + U_3$$

$$= \frac{V^2}{2(EI)_2} \left[\frac{L^3}{3} - L^2 L_1 + LL_1^2 - \frac{L_1^3}{3} \right] + \frac{V^2}{2(EI)_1} \left[L^2 L_1 - LL_1^2 + \frac{L_1^3}{3} \right]$$

$$= \frac{V^2}{2(EI)_2} \frac{L^3}{3} - \frac{V^2}{2} \left[L^2 L_1 - LL_1^2 + \frac{L_1^3}{3} \right] \left[\frac{1}{(EI)_2} - \frac{1}{(EI)_1} \right] \quad (4.55)$$

which, using

$$L_1 = L - L_2 - a$$

gives

$$U = \frac{V^2}{2(EI)_2} \frac{L^3}{3} - \frac{V^2}{2} \left[L^2(L - L_2 - a) - L(L - L_2 - a)^2 + \frac{(L - L_2 - a)^3}{3} \right]$$

$$\cdot \left[\frac{1}{(EI)_2} - \frac{1}{(EI)_1} \right] \quad (4.56)$$

Then, using Equation 4.21, we get

$$G = -\frac{\partial U}{\partial A} = -\frac{\partial U}{w \partial a}$$

$$= \frac{1}{w} \left(-\frac{V^2}{2} \right) \left[-L^2 + 2L(L - L_2 - a) - (L - L_2 - a)^2 \right] \left[\frac{1}{(EI)_2} - \frac{1}{(EI)_1} \right]$$

$$= \frac{V^2}{2w} \left[\frac{1}{(EI)_2} - \frac{1}{(EI)_1} \right] (L_2 + a)^2 = \frac{V^2(L_2 + a)^2}{2w} \left[\frac{1}{(EI)_2} - \frac{1}{(EI)_1} \right] \quad (4.57)$$

where w is the width of the beam perpendicular to the page in Figure 4.21. Note that an additional minus sign is introduced in this equation to account for the fact that the delamination area A is increasing for decreasing x. The bending stiffnesses $(EI)_1$ and $(EI)_2$ are determined using standard equivalent properties for composite beams [12].

The case where instead of an applied shear force V a bending moment M is applied can also be solved in an analogous manner. In this particular case, the bending moment in the skin is equal to M for $L_1 \leq x \leq L$. For the skin and flange portion, with $x < L_1$, the bending moment is also equal to M. The corresponding strain energies are

$$U_3 = \int_{L_1+a}^{L} \frac{M^2}{2(EI)_2} dx = \frac{M^2}{2(EI)_2}(L - L_1 - a)$$

$$U_2 = \int_{L_1}^{L_1+a} \frac{M^2}{2(EI)_2} dx = \frac{M^2}{2(EI)_2}(L_1 + a - L_1) = \frac{M^2 a}{2(EI)_2}$$

$$U_1 = \int_{0}^{L_1} \frac{M^2}{2(EI)_1} dx = \frac{M^2 L_1}{2(EI)_1} \qquad \text{(4.58a–c)}$$

Combining Equations 4.58a–c, we get the total U as

$$U = U_1 + U_2 + U_3$$

$$= \frac{M^2}{2(EI)_2}(L - L_1 - a) + \frac{M^2 a}{2(EI)_2} + \frac{M^2 L_1}{2(EI)_1} = \frac{M^2}{2(EI)_2}(L - L_1) + \frac{M^2 L_1}{2(EI)_1}$$

$$= \frac{M^2 L}{2(EI)_2} + \frac{M^2 L_1}{2}\left(\frac{1}{(EI)_1} - \frac{1}{(EI)_2}\right) \qquad \text{(4.59)}$$

which, using

$$L_1 = L - L_2 - a$$

gives

$$U = \frac{M^2 L}{2(EI)_2} + \frac{M^2(L - L_2 - a)}{2}\left(\frac{1}{(EI)_1} - \frac{1}{(EI)_2}\right)$$

$$= -\frac{M^2(L_2 + a)}{2}\left(\frac{1}{(EI)_1} - \frac{1}{(EI)_2}\right) + \frac{M^2 L}{2(EI)_1} \qquad \text{(4.60)}$$

Then, using Equation 4.21, we get

$$G = -\frac{\partial U}{\partial A} = -\frac{\partial U}{w \partial a} = \frac{M^2}{2}\left(-\frac{1}{(EI)_1} + \frac{1}{(EI)_2}\right) \qquad \text{(4.61)}$$

When both a shear force and a bending moment are applied, the total strain energy release rate can be obtained by adding Equations 4.57 and 4.62.

The above results in Equations 4.57 and 4.62 can be used to derive the energy release rate expression for a couple of important special cases. Considering Figure 4.21 and

Equation 4.57, it can be seen that, when $a = 0$, the energy release rate takes a finite value:

$$G_{onset} = \frac{V^2 L_2^2}{2w} \left[\frac{1}{(EI)_2} - \frac{1}{(EI)_1} \right] \tag{4.62}$$

This can be viewed as the value of G at onset of delamination. And to the extent that Figure 4.21 can be seen as representing a section of a stiffener flange attached to a skin, Equation 4.62 can be used to predict onset of delamination under a transverse shear load V. It is important to note that the discussion and assumptions of the last paragraph of Section 4.3.4 carry over here also. Assuming that the limiting value of Equation 4.62 does not violate any of the assumptions already made, one can equate the right-hand side to an experimentally determined critical energy release rate G_{onset} to determine when a delamination between skin and stiffener would occur.

Another special case of interest is the situation where the configuration of Figure 4.21 is mirrored with respect to a horizontal plane to obtain a double cantilever beam (DCB), as shown in Figure 4.22.

Comparing Figures 4.21 and 4.22, one can see that $L_2 + a$ from Figure 4.21 is the same as a in Figure 4.22. Also, region 1 is no longer present. Now if the two beams (or the skin and flange) have the same geometry and stiffness, one can write $(EI)_2 = EI$ for simplicity. Finally, multiplying the result of Equation 4.57 by 2 to account for the second beam, which doubles the energy produced when the crack grows, gives

$$G_I = 2\frac{V^2 a^2}{2w} \left[\frac{1}{(EI)} \right] = \frac{V^2 a^2}{wEI} \tag{4.63}$$

with w the width perpendicular to the page of Figure 4.22 and EI the stiffness of each of the two beams with respect to its own neutral axis.

It can be seen from Equation 4.63 that, as the delamination size a tends to zero, G also tends to zero. This is unlike the previous results of Equations 4.42a and 4.62 where G is finite when $a = 0$. It should also be pointed out here that the expression in Equation 4.63 for the DCB accounts only for the bending energy. For deep beams where transverse shear effects are significant, a correction to that equation is necessary (see Section 4.3.6). Also, for unsymmetric beams, an approximate expression can be

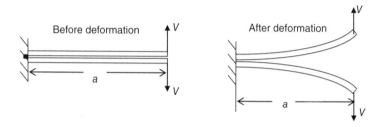

Figure 4.22   Double cantilever beam of length a

obtained if $w(EI)$ in Equation 4.63 is replaced by $\overline{D}_{11}$, the 11 entry of the reduced bending stiffness matrix of the laminate making up each beam (see Equation 4.3).

4.3.6 Double Cantilever Beam and End Notch Flexure Specimen

Of particular importance for composite material characterisation when it comes to delamination growth are the DCB and the end notch flexure (ENF) specimens.

4.3.6.1 Double Cantilever Beam (DCB)

This was examined briefly as a special case in Section 4.3.5. It is very useful in material characterisation because it can be used to determine G_{IC}, the critical energy release rate in mode I (see, for example, Figure 4.18). Here, an alternate way to derive the result of Equation 4.63 is provided. The reason for this relates to the interchange between the internal potential energy and the work done, both of which can be used in determining the strain energy release rate as was mentioned in Section 4.3.4. Consider the situation to the right of Figure 4.22. Under constant load, the total work done by the applied forces V equals the internal strain energy. The tip deflection of a cantilever beam is given by

$$\delta_{tip} = \frac{Va^3}{3EI} \tag{4.64}$$

Then the work done by one of the forces V is

$$W_1 = \frac{1}{2}V\delta_{tip} = \frac{V^2a^3}{6EI} \tag{4.65}$$

and, for both forces in Figure 4.22 acting, the total work W is given by

$$W = 2W_1 = \frac{V^2a^3}{3EI} \tag{4.66}$$

Now the energy release rate is obtained by recognising that the strain energy U and the work done are equal and that the change in strain energy when the delamination advances is the negative of the work done (see also Equation 4.26). Therefore

$$G = -\frac{\partial U}{\partial A} = \frac{\partial W}{\partial A} = \frac{1}{w}\frac{\partial W}{\partial a} = \frac{V^2a^2}{wEI} \tag{4.67}$$

where Equation 4.66 was used. This is the same result as in Equation 4.63.

An approach often used in determining the energy release rate makes use of the compliance C. This approach is useful in reducing test data, as compliance is the constant of proportionality that relates force to displacement. For a DCB, the compliance relating the total opening displacement $2\delta_{tip}$ is found from Equation 4.64:

$$C = 2\frac{a^3}{3EI} \tag{4.68}$$

For a linear system, force and deflection are related via the compliance:

$$\delta = CV \qquad (4.69)$$

and thus, the complementary energy, which equals the internal energy U, can be written as

$$U = \frac{1}{2}V\delta = \frac{1}{2}CV^2 \qquad (4.70)$$

where Equation 4.69 was used.

Under constant force, the energy release rate can be written as

$$G = -\frac{\partial U}{\partial A} = -\frac{1}{2}V^2\frac{\partial C}{\partial A} = -\frac{V^2}{2w}\frac{\partial C}{\partial a} \qquad (4.71)$$

and using Equation 4.68, we get

$$G = \frac{V^2 a^2}{wEI}$$

which is the same result as in Equations 4.63 and 4.67. Note that the minus sign of Equation 4.71 does not show up in this result because the compliance C decreases as a increases and thus $\partial C/\partial a$ is negative. Equation 4.71 is used in tests where the derivative $\partial C/\partial a$ is measured experimentally. The use of the DCB test to obtain G_{IC} is described in [15].

As already mentioned in the previous section, Equations 4.67 and 4.63 are a first (and for long beams very good) approximation to the energy release rate for a DCB. Corrections typically result in effectively increasing the beam length by a small amount that accounts for end rotations [16].

Equation 4.63 can be used to predict delamination growth. A plot of G as a function of delamination length for various applied loads V_1, V_2, and so on, is shown in Figure 4.23. Focusing on the left of Figure 4.23, suppose the delamination length is a_1. In a stroke-controlled test (applied displacement), there will be no growth of this

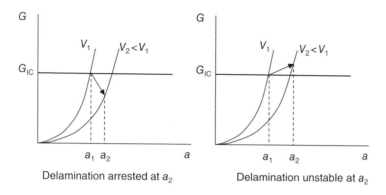

Figure 4.23 Use of energy release rate to determine delamination growth

delamination until the energy release rate G equals the critical energy release rate G_{IC}, which is a material constant and is shown as a horizontal line in Figure 4.23. When growth starts, the applied load is V_1 because the G versus a curve which corresponds to a_1 when $G = G_{IC}$ has applied load equal to V_1. Suppose the delamination grows to a new length a_2. During growth, the displacement is controlled and is kept constant. This means that the applied load is reduced from V_1 to V_2. Depending on the value of a_2, the specimen may follow the arrow shown to the left of Figure 4.23 and end up at a G value lower than G_{IC}. In such a case, the delamination will stop. If the scenario shown to the right of Figure 4.23 is followed, the new G value is greater than G_{IC} and the delamination will grow further. Whether the delamination will grow unstably or will be arrested will depend on the dynamics of the test and whether G_{IC} is exceeded for any new delamination length.

Obviously, to delay delamination onset and growth, tough materials with high G_c in the respective modes are preferred.

4.3.6.2 End Notch Flexure Specimen (ENF)

This is a situation where three-point bending is used to induce pure shear in a delaminated specimen, as shown in Figure 4.24. The general case where the delamination of length a can be anywhere through the thickness of the beam will be considered first.

The beam is divided into three regions as shown. From standard beam theory, it can be shown that the deflection w_1 in Region 1 is given by

$$w_1 = -\frac{P}{12EI_1}x_1^3 + C_1 x_1, \quad 0 \le x_1 \le \frac{L}{2}$$

$$w_1 = -\frac{P}{2EI_1}\left(\frac{Lx_1^2}{2} - \frac{x_1^3}{6}\right) + \frac{PL^2}{8EI_1}x_1 + C_1 x_1 - \frac{PL^3}{48EI_1}, \quad \frac{L}{2} \le x_1 \le L - a \quad (4.72)$$

where C_1 is an unknown constant. Physically, C_1 is the rotation of the beam at the right end, $x_1 = 0$. Note that, because of the presence of the delamination, the rotations at the two ends of the beam will not be the same.

For convenience, define a new coordinate x_2 in region 2 such that $x_2 = 0$ at the delamination tip (when $x_1 = L - a$). Let M_2 the bending moment in region 2 and M_3

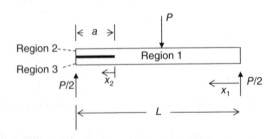

Figure 4.24   End notch flexure specimen under load

in region 3. Now the radius of curvature of Region 2 and that of Region 3 are related to the radius of curvature R or Region 1 via the following relationships:

$$R_2 = R - \frac{2h - t_2}{2}$$

$$R_3 = R + \frac{2h - t_3}{2}$$

with t_2 and t_3 the thicknesses of Region 2 and Region 3. Obviously, if deflections are small and the beam is thin, $R \approx R_2 \approx R_3$.

Now from beam theory

$$M = \frac{EI}{R}$$

which, combined with the fact that $M_2 + M_3 = M_1$, leads to

$$M_2 = \frac{R + ((2h - t_3)/2)}{(R - ((2h - t_2)/2))EI_3 + (R + ((2h - t_3)/2))EI_2} EI_2 M_1 = r_1 EI_2 M_1$$

$$M_3 = \frac{R - ((2h - t_2)/2)}{(R - ((2h - t_2))/2)EI_3 + (R + ((2h - t_3)/2))EI_2} EI_3 M_1 = r_2 EI_3 M_1$$

where r_1 and r_2 are introduced for convenience to save carrying through the long fractions in the right-hand sides.

The displacement w_2 in Region 2 can now be determined from beam theory and by matching the deflection and slope at $x_2 = 0$ with those from Region 1. The result is

$$w_2 = -\frac{Pr_1}{2}\left(a\frac{x_2^2}{2} - \frac{x_2^3}{6}\right) - \frac{P}{4EI_1}(L^2 - a^2)x_2 + \frac{PL^2}{8EI_1}x_2 + C_1 x_2$$

$$-\frac{P}{12EI_1}(L - a)^2(2L + a) + \frac{PL^2}{8EI_1}(L - a) + C_1(L - a) - \frac{PL^3}{48EI} \quad (4.73)$$

The unknown constant C_1 can now be determined by requiring that w_2 be zero at $x_2 = a$. This gives

$$C_1 = \frac{1}{L}\left[\frac{Pr_1 a^3}{6} - \frac{P}{4EI_1}\left(-\frac{L^3}{4} + \frac{2}{3}a^3\right)\right]$$

This allows the determination of the vertical deflection at $x_1 = L/2$, the mid-point of the entire beam:

$$w_1\left(x_1 = \frac{L}{2}\right) = -\frac{PL^3}{96EI_1} + \frac{1}{2}\left[\frac{Pr_1 a^3}{6} - \frac{P}{4EI_1}\left(-\frac{L^3}{4} + \frac{2}{3}a^3\right)\right] \quad (4.74)$$

The work done W when load P is applied can now be obtained as half the product of the load and the corresponding displacement:

$$W = \frac{1}{2}Pw\left(x_1 = \frac{L}{2}\right) = -\frac{P^2 L^3}{192EI_1} + \frac{P^2 r_1 a^3}{24} - \frac{P^2}{16EI_1}\left(-\frac{L^3}{4} + \frac{2}{3}a^3\right)$$

Now, for a linear system the work done equals the internal strain energy. Therefore, using Equation 4.21 and accounting for the fact that as strain energy is released positive work is done,

$$G = \frac{1}{w} \frac{\partial W}{\partial a} = \frac{P^2 a^2}{8b} \left(r_1 - \frac{1}{EI_1} \right) \tag{4.75a}$$

The case of the ENF specimen with a delamination at the mid-plane can now be recovered as a special case of Equation 4.75a. In this case, and assuming the radii of curvature of the three regions are large compared to the region thicknesses, $r_1 = 1/(EI_2 + EI_3)$ and $EI_2 = EI_3 = EI_1/8$, one can substitute in Equation 4.75a to obtain

$$G = \frac{3}{8} \frac{P^2 a^2}{bEI_1} = \frac{9}{16} \frac{P^2 a^2}{b^2 h^3} \tag{4.75b}$$

where the fact that $I_1 = b(2h)^3/12$ was used.

The ENF test can be used to determine the critical energy release rate G_{IIC}. Unlike the DCB test for which an ASTM (American Society for Testing and Materials) standard is available [15], there is no ASTM standard for the ENF test. Proposed test procedures for the ENF test can be found in [17]. As for the case of the DCB test, the result presented here, Equation 4.75a, is a first-order approximation valid for relatively long and slender beams where root rotations are negligible. A more accurate expression can be found in [18].

It should be noted that the discussion here is more of an overview providing starting points for the analysis of composite structures with delaminations. More detailed discussion can be found in the literature, for example, [19–21].

4.3.7   The Crack Closure Method

The analytical solutions presented in Sections 4.3.4–4.3.6 belong to a minority of situations where the strain energy release rate can be computed analytically. In general composite structures, the geometry and loading are such that it is not possible to resort to such relatively simple expressions. Usually, some form of numerical method is necessary.

In an attempt to relate far-field loading to local energy release rate values, Williams [22] followed by Schapery and Davidson [23] used the classical laminated plate theory to relate local loads at the delamination front – bending moments, normal and shear loads – to the total energy release rate, and even split it to its mode I and II components. Knowledge of the 'mode mixity' is very important when more than one mode is present. The growth criterion to be used is a function of the relative magnitudes of G_I, G_{II} and G_{III}.

This approach is powerful but still requires a way of relating far-field applied loads to the local loads. For realistic composite structures, this means the use of finite elements. Once the finite element model has been constructed, using it to determine the energy release rate via the crack closure method is an alternative which is very often used in practice.

The crack closure technique evaluates numerically the derivatives in Equation 4.21 or 4.27. It amounts to placing a delamination of length a in the structure and letting it grow to a length $a + \Delta a$. Calculating the strain energy for the two cases, subtracting them and dividing by Δa would give $\Delta U/\Delta a$, which, in the limit as $\Delta a \to 0$, would give the energy release rate G. There are a couple of interesting points or issues with this approach. The first is that the direction in which the delamination will grow must be known. As long as the delamination stays between two specific plies, this may not be too challenging even though one still would have to trace the delamination front to find which way a fully embedded delamination, such as an elliptical delamination, would grow. This, for example, would be the case in Section 4.3.2. The second is a matter of computational efficiency, as two separate finite element runs are required. To circumvent the latter issue, the virtual crack closure technique (VCCT) has been implemented, which requires only one finite element run to accomplish the determination of the energy release rate.

As the stress field in front of the delamination front is singular, within the context of linear anisotropic elasticity with homogenised matrix and fibre properties in each ply, finite element methods using the VCCT or the crack closure technique suffer often from numerical problems such as oscillations near the crack tip caused by interpenetration of the crack surfaces. A promising method to eliminate such numerical problems has been proposed by Zou *et al.* [24], where stress resultants and derivatives of the displacements were used to obtain the energy release rates for each individual mode. An excellent overview and discussion of the development and evolution of the VCCT and different modelling techniques is provided by Krüger [25].

Here, before giving the pertinent equations for determining G using VCCT, it is worthwhile revisiting the single cantilever beam specimen. The reason is that solving this problem in a different way that parallels the VCCT will make it easier to understand the VCCT. The situation is shown in Figure 4.25. A cantilever beam of length $a + \Delta a$ and bending stiffness EI is under a tip load P. A load F is exerted at $x = \Delta a$. First, the magnitude of F is required that would make the beam deflection at $x = \Delta a$ equal to zero. This would be equivalent to closing the crack and reducing its length from $a + \Delta a$ to a.

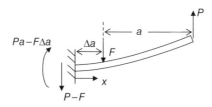

Figure 4.25   Cantilever beam: Setting the deflection at Δa equal to zero

The reactions, bending moment and shear force are as shown in Figure 4.25. The deflection $\delta(x)$ at any point x along the beam *if only P is acting* is given by standard beam theory:

$$\delta(x) = \frac{P}{EI}\left((a + \Delta a)\frac{x^2}{2} - \frac{x^3}{6}\right) \tag{4.76}$$

Then, the deflection at $x = \Delta a$ is given by

$$\delta(\Delta a) = \frac{P}{EI}\left((a + \Delta a)\frac{\Delta a^2}{2} - \frac{\Delta a^3}{6}\right) \tag{4.77}$$

and the tip deflection at $x = a + \Delta a$ is given by

$$\delta(a + \Delta a) = \frac{P(a + \Delta a)^3}{3EI} \tag{4.78}$$

We determine now F such that $\delta(a + \Delta a) = 0$. This is done using Castigliano's second theorem. The deflection in the direction of force F is given by

$$\delta(\Delta a) = \frac{\partial U}{\partial F} = \frac{\partial}{\partial F}\left[\frac{1}{2}\int_a^{a+\Delta a}\frac{M^2}{EI}dx\right] \tag{4.79}$$

where $M(x)$, the bending moment in the beam, given by

$$M(x) = P(a + \Delta a) - F\Delta a - (P - F)x, \quad 0 \le x \le \Delta a$$
$$= P(a + \Delta a) - Px, \quad \Delta a < x \le a + \Delta a \tag{4.80}$$

Placing Equation 4.80 into Equation 4.79, evaluating the integrals, setting $\delta(\Delta a) = 0$ and solving for F gives

$$F = P\left(\frac{3}{2}\frac{a + \Delta a}{\Delta a} - \frac{1}{2}\right) \tag{4.81}$$

Therefore, the work done by force F in closing the crack, that is, making the deflection to change from its value in Equation 4.77 to zero, is

$$\Delta W = \frac{1}{2}F\delta(\Delta a) = \frac{P^2}{24EI}\Delta a(3a + 2(\Delta a)^2) \tag{4.82}$$

At the same time, the internal energy U in the beam, after F is applied, is given by the quantity in brackets on the right-hand side of Equation 4.79. Evaluating, after considerable algebra, gives

$$U_a = \frac{1}{2EI}\left[\frac{F\Delta a^3}{3} - 2P^2\Delta a^3 - \frac{2}{3}PF\Delta a^3 - PFa\Delta a^2 - P^2a\Delta a^2 + P^2\frac{a^3}{3} + P^2a^2\Delta a\right] \tag{4.83}$$

The internal energy before F is applied is obtained in an analogous manner (with $F = 0$):

$$U_{a+\Delta a} = \frac{1}{2EI}\frac{P^2(a + \Delta a)^3}{3} \tag{4.84}$$

In order to compute the energy release rate G, it should be recognised here that the force F is not constant. It starts from zero before the crack is closed and increases to its maximum value given by Equation 4.81 after the crack is closed. Thus, Equations 4.28 and 4.29 must be used. The derivative of U with respect to the crack surface area created is given by

$$-\frac{\partial U}{\partial A} = \lim_{\Delta a \to 0} \left[-\frac{U_a - U_{a+\Delta a}}{w\Delta a} \right]$$

$$= \frac{1}{2EI} \lim_{\Delta a \to 0} \left[F^2 \frac{\Delta a^2}{3} - 2PF\frac{\Delta a^2}{3} - PFa\Delta a - P^2 a\Delta a - 2P^2 \Delta a^2 + P^2 a^2 \right] \quad (4.85)$$

Similarly, from Equation 4.82 we have

$$\frac{\partial W}{\partial A} = \lim_{\Delta a \to 0} \left[\frac{\Delta W}{w\Delta a} \right] = \lim_{\Delta a \to 0} \frac{P^2}{24wEI}(3a + 2(\Delta a)^2) \quad (4.86)$$

Therefore, applying Equation 4.28, using Equation 4.81 and evaluating the limits in Equations 4.85 and 4.86 gives the final expression for G:

$$G = \lim_{\Delta a \to 0} \frac{1}{2wEI} \left[\left(\frac{P^2 a^2}{4} - 4P^2 a\Delta a - 2P^2 \Delta a^2 \right) + \left(\frac{1}{12}(3a + 2\Delta a)^2 \right) \right] = \frac{P^2 a^2}{2wEI} \quad (4.87)$$

which, if multiplied by 2 to account for two beams, is exactly the same as the result obtained earlier for the DCB in Equation 4.63.

The VCCT is based on the premise that the energy released when the crack extends by Δa is exactly equal to the energy required to close the crack and reduce its length by the same amount. Therefore, the previous derivation is directly applicable here. The work done by force F to close the crack provided the needed contribution for obtaining the energy release rate.

Taking the previous derivation one step further, consider now the local situation in a finite element model near a delamination front. For simplicity, assume this is a two-dimensional problem, as shown in Figure 4.26.

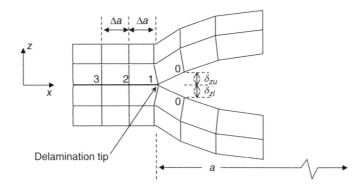

Figure 4.26   Finite element mesh in the vicinity of a delamination tip

If the crack in Figure 4.26 extends from a to $a + \Delta a$, opening node 1, the displacement is assumed to be the same as if a crack of length $a + \Delta a$ extended to $a + 2\Delta a$, opening node 2. At the same time, the energy required to close the crack of length $a + 2\Delta a$ at node 2 is assumed to be the same as that required to close the crack $a + \Delta a$ at node 1. These assumptions will be valid for sufficiently small values of Δa. Then, with reference to the situation of Figure 4.26, with the crack stopping at node a, the displacement δ ahead of that node at node 0 will be approximately the same as the displacement at node 1 when node 1 is released. Also, the force F_z closing node 1 is approximately the same as that required to close node 0. Thus, the energy release rate can be approximated by

$$G_I \simeq \frac{\Delta U}{\Delta A} = \frac{1}{2} \left(\frac{F_z \left(\delta_{zu} - \delta_{zl} \right)}{w \Delta a} \right) \tag{4.88}$$

In the limit as $\Delta a \to 0$, Equation 4.88 will be exact.

Note that the 1/2 factor in Equation 4.88 comes from the integration of $dW = F_z d\delta$, where the force and displacement are linearly related. For a two-dimensional problem using four-noded elements, w is the width of the structure perpendicular to the page of Figure 4.26 and Δa is the horizontal distance between successive nodes 0 and 1. Equation 4.88 gives the mode I energy release rate. For mode II, a similar expression can be used but involving the horizontal force and the horizontal displacements δ_x for each node. An analogous extension can be made for mode III, but in this case solid elements must be used. Expressions for different types of elements, shell or solid and modes, I, II or III, can be found in [25].

One final comment related to the size of Δa and potential mesh dependence of the results is in order. As already suggested, infinite stresses at the delamination tip may lead to oscillations of displacements near the delamination tip. Therefore, as Δa changes, the deflections δ_{zu} and δ_{zl} may vary substantially, causing convergence problems. Care must be exercised in selecting the proper mesh, or the approach in [24] is recommended.

The problem of oscillations is quite pronounced at bi-material interfaces where the two materials are different orthotropic materials. This is the case of a sandwich structure where the facesheet and core have drastically different properties. Modelling a disbond, that is, a delamination between the facesheet and core, requires special attention and is not dealt here. The reader is referred to specialised references such as the work by Carlsson and Kardomateas [26].

4.4 Strength of Materials Versus Fracture Mechanics – Use of Cohesive Elements

In most discussions about delaminations so far, the delamination was assumed to be present already in the structure and the possibility of its growth was examined through buckling or energy release rate calculations. Onset of delamination was only addressed as a special case of some problems where, upon letting the delamination size go to zero, the energy release rate took a finite nonzero value. Under certain circumstances, as mentioned in Sections 4.3.4 and 4.3.5, this value of G can be used to predict the onset of delamination by setting it equal to the critical energy release rate for delamination onset, which is determined experimentally.

It is important to note that, as far as onset of delamination is concerned, it is also possible to use a strength-of-materials approach to determine it. If stresses, including out-of-plane (inter-laminar) stresses, can be calculated, they can be used in a failure criterion to predict onset of delamination. For this purpose, besides finite elements, methods such as those presented in [27] for skin–stiffener separation can be used.

Given that the two approaches are quite different, it is worth comparing their predictions. For the strength-of-materials approach, the method presented by Kassapoglou and Lagacé [28] was used by Brewer and Lagacé in [29] in conjunction with a quadratic delamination criterion (QDC) from [29] to predict the onset of delamination. The QDC has the form [29]

$$\left(\frac{\overline{\tau_{xz}}}{Z_{xz}}\right)^2 + \left(\frac{\overline{\tau_{yz}}}{Z_{yz}}\right)^2 + \left(\frac{\overline{\sigma_z}}{Z_t}\right)^2 = 1 \qquad (4.89)$$

where z is the out-of-plane direction and the overbar over the inter-laminar stresses denotes that the stresses are averaged over a characteristic distance x_{avg}. Thus

$$\overline{\sigma} = \frac{1}{x_{avg}} \int_0^{x_{avg}} \sigma \, dx \qquad (4.90)$$

Z_{xz}, Z_{yz} and Z_t are out-of-plane shear and normal strength values. If the inter-laminar normal stress σ_z is compressive at the interface of interest, then the last term in Equation 4.89 is omitted. This is conservative because out-of-plane compression tends to delay delamination onset caused by inter-laminar shear.

The use of the averaging distance is analogous to the Whitney–Nuismer approach for predicting failure in laminates with holes (see Section 2.4). In a sense, the averaging distance recognises that damage will start but will not necessarily lead to final failure and will create a damage zone which will keep the inter-laminar stresses from reaching their maximum values predicted by an anisotropic elasticity solution. The value of x_{avg} is determined experimentally [29] and is not the same as the averaging distance for holes in Sections 2.4 or 2.6.

For the energy release rate predictions, the critical energy release rate G_c was back-calculated in [29] using the approach of Section 4.3.4 and Equation 4.43a to match a sub-set of the test results. Then, this value of G_c was used to make additional predictions for different specimens. Results comparing the strength-of-materials (QDC) and energy release rate (G_c) predictions taken from [29] are shown in Figures 4.27–4.29. During the tests, the number n of identical plies next to each other was varied.

It can be seen from Figures 4.27–4.29 that the two methods have nearly the same accuracy. The strength-of-materials method (QDC) seems to do better for the

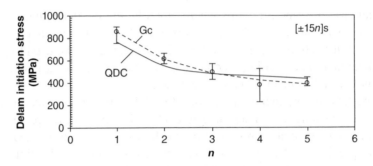

Figure 4.27 Onset of delamination predictions by strength-of-materials and fracture mechanics approaches compared to test results for [±15*n*]s laminate.

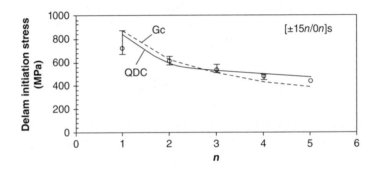

Figure 4.28 Onset of delamination predictions by strength-of-materials and fracture mechanics approaches compared to test results for [±15*n*/0*n*]s laminate

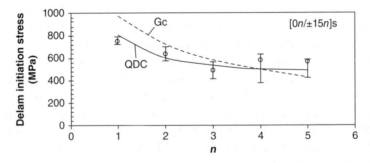

Figure 4.29 Onset of delamination predictions by strength-of-materials and fracture mechanics approaches compared to test results for [0*n*/±15*n*]s laminate

[0*n*/±15*n*]s laminate in Figure 4.29, while the two methods are equally effective for the other two laminates in Figures 4.27 and 4.28. It should be emphasised, however, that both methods require the experimental determination of model parameters or material properties. The strength-of-materials method requires determination of the averaging distance x_{avg} and the values Z_{xz}, Z_{yz} and Z_t, some of which (Z_t and Z_{yz}) are hard to get when the fibres are in the x-direction. The fracture mechanics approach requires determination of G_c for the particular laminate and ply interface of interest. As an additional issue, the fracture mechanics approach of Equation 4.43a requires knowledge of the sub-laminate stiffnesses E_i, which, as discussed in Section 4.3.4, are dependent on loading, boundary conditions and sub-laminate lay-up. The next section briefly discusses an approach that combines the strength of materials and fracture mechanics in one model.

4.4.1   Use of Cohesive Elements

One important and effective approach that can model both onset of delamination and delamination growth is based on the use of cohesive elements. This is done within a finite element model where zero-thickness cohesive elements are located at the interfaces of interest. Each element has a constitutive law that describes how the interface stresses or tractions relate to local separation displacements. Two examples of cohesive laws are given in Figure 4.30.

The initial linear portion of the cohesive law represents the standard linear stress–strain curve to onset of failure which occurs at separation δ_o when the interface strength is reached. This is not final failure which occurs when the displacement between the two interfaces jumps to δ_f. The portion of the curve after the interface strength is reached, between δ_o and δ_f, represents stiffness degradation. Obviously, knowing the exact form in which stiffness degrades after initial failure will define the shape (linear, exponential or other) in Figure 4.30 for $\delta > \delta_o$. One piece of information that helps in fixing this shape is that it can be shown from fracture mechanics considerations and use of the J-integral that the area under the entire curve must equal the critical energy release rate G_c for the specific mode of interest. This fixes completely the shape of the bi-linear law on the left of Figure 4.30 but is not sufficient to fix the exponential law on the right. Additional information or assumptions are necessary. The combination of the interfacial strength and energy

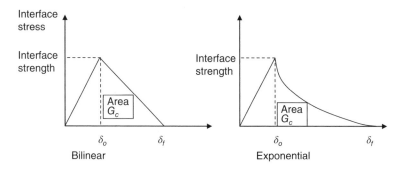

Figure 4.30   Cohesive laws

release rate in one cohesive law brings together the strength-of-materials and the fracture mechanics approaches.

Cohesive elements are a powerful tool that can be used to predict both delamination onset and its growth. Depending on the situation, the finite element model may be very intensive computationally. Also, in some cases, the answer may depend on the mesh size because the cohesive elements are placed in between 'standard' elements and growth of a crack can only be along the paths that the interfaces of the standard elements will provide. The reader is referred to the literature, for example, [30–33], for more details.

Exercises

4.1 The coin-tapping method used in a factory can reliably detect delaminations of length at least 38 mm and width 34.5 mm. A skin panel is made with lay-up [45/−45/0$_3$/45/−45/90]s with the following material properties:

$$E_x = 137.9\,\text{GPa}$$
$$Ey = 11.72\,\text{GPa}$$
$$v_{xy} = 0.29$$
$$G_{xy} = 5.17\,\text{GPa}$$
$$t_{ply} = 0.1524\,\text{mm}$$
$$X_t = 2068\,\text{MPa}$$
$$X_c = 1379\,\text{MPa}$$
$$Y_t = 68.94\,\text{MPa}$$
$$Y_c = 303.3\,\text{MPa}$$
$$S = 124.1\,\text{MPa}$$

The dimension of the skin panel is 558.8 mm × 431.8 mm. It is compression-critical, and the compression load is along the long dimension.

(a) For an elliptical delamination between the second and third plies and aspect ratio of 1.1 (where longer dimension is along the compression load), determine the size beyond which the delamination drives failure. Repeat for a delamination between fifth and sixth plies. Show your solution graphically.

(b) Given your answers in part (a), is the coin-tapping method sufficient for delamination detection?

(c) What happens if the aspect ratio is 2 instead of 1.1? (do not repeat the entire solution; just find what happens to the delaminations).

Note that in this problem you cannot neglect D_{16} and D_{26} unless they are smaller than 17% of the next largest entry in the D matrix.

4.2 (a) A typical eight-ply Gr/Epoxy quasi-isotropic lay-up is to be tested in tension. The lay-up consists only of 45°, −45°, 0° and 90° plies. There is a concern that the laminate may fail by edge delamination before it reaches the material strength (first ply failure) load. See Figure E4.1 below.

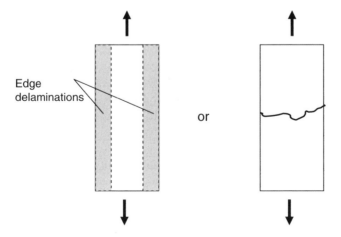

Figure E4.1 Edge delamination versus material strength failure for a laminate

The following four variants of the basic layup are considered: (i) [45/−45/0/90]s, (ii) [45/−45/90/0]s, (iii) [45/0/−45/90]s and (iv) [45/90/−45/0]s. Assume that the material used has the following properties:

Unidirectional tape Gr/Epoxy
$E_x = 131$ GPa
$E_y = 11.4$ GPa
$v_{xy} = 0.31$
$G_{xy} = 5.17$ GPa
$t_{\text{ply}} = 0.3048$ mm

Determine which of the four lay-ups is most likely to have edge delamination and where (at which ply interface). Determine which of them is least likely and its critical ply interface. Assume that the mid-plane, which always has two plies of the same orientation on either side (which means the resin layer between plies is very small, if it exists at all), is unlikely to delaminate.

(b) If the strength properties are as given below, determine which of the lay-ups above will fail by material strength and not by edge delamination.

$X_t = 2068$ MPa
$X_c = 1723$ MPa
$Y_t = 68.9$ MPa
$Y_c = 303.3$ MPa
$S = 124.1$ MPa
$G_c = 112.9$ J m^{-2}

4.3 A wing skin is made out of a sandwich with 6.35-mm-thick core and facesheet lay-up: $[45/-45/0_4/90]$s. The material properties for the facesheet material are as follows:

$$
\begin{aligned}
E_x &= 137.9 \, \text{GPa} \\
E_y &= 11.7 \, \text{GPa} \\
v_{xy} &= 0.31 \\
G_{xy} &= 4.82 \, \text{GPa} \\
t_{\text{ply}} &= 0.1524 \, \text{mm}
\end{aligned}
$$

A specific panel of the wing skin is under compression, and its length is fixed at 508 mm (applied compression is along the length dimension). Its width has not been decided yet and is allowed to vary between 100 and 350 mm. Whatever the final width, the resulting panel will have some form or reinforcement and/or attachment all around which will impose simply-supported boundary conditions at the panel edges. The panel is buckling-critical (to simplify the problem, neglect all other failure modes).

As the manufacturing process is not perfect, elliptical delaminations may occur at any of the interfaces: 45/−45, −45/0, 0/90, 90/0 or 0/−45. Only one delamination at a time is likely (i.e., no multiple delaminations at different interfaces).

(a) By treating the delamination boundaries as clamped, and assuming that the aspect ratio b/a equals 0.3, 1 or 1.5, obtain three charts (one for each value of b/a) that show the facesheet load per panel width that causes the delamination to buckle as a function of delamination half length a (Figure E4.2).

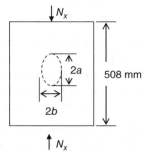

Figure E4.2   Embedded elliptical delamination in a laminate under compression

On each chart, determine the minimum delamination size for each of the delamination locations mentioned above for different panel widths (start from 100 mm and go to 350 mm width in 50-mm increments)

(b) Design chart creation. Create a chart where the x-axis has the delamination half-length and the y-axis the delamination aspect ratio. For this chart, the delamination is assumed to be at the 45/−45 interface only. The chart should show curves of constant delamination buckling load (per facesheet). Show only three curves, one for 437.5 N mm^{-1}, one for 934.5 N mm^{-1} and one for 1312.5 N mm^{-1}. Design charts of this kind can be very useful in day-to-day design, as they can quickly give an idea of the load capability given a damage size or the damage size given the load requirement.

(c) Tie in the inspection capability. Assume that, based on work in the previous parts of this assignment (and other considerations), the width of the panel is set to 250 mm. To minimise cost, the inspection method chosen is coin tap. Based on previous experience, the coin-tap method can reliably detect a delamination or area equal to the area of a circle with diameter 12 mm (but the delamination to be detected does not have to be circular; it can be elliptical and its area can be approximated by πab with a and b as defined above). Superimpose on the chart of part (b) above a single curve that shows the minimum delamination inspection capability. What is the minimum delamination length that a designer should design to?

References

[1] Kassapoglou, C. (2013) *Design and Analysis of Composite Structures*, 2nd edn, Chapter 10, John Wiley & Sons, Inc., New York.

[2] Kardomateas, G.A. and Schmueser, D.W. (1988) Buckling and Postbuckling of delaminated composites under compressive loads including transverse shear effects. *AIAA J.*, **26** (3), 337–343.

[3] Kassapoglou, C. (2013) *Design and Analysis of Composite Structures*, 2nd edn, chapter 3.3, John Wiley & Sons, Inc, New York.

[4] Kassapoglou, C. (2013) *Design and Analysis of Composite Structures*, 2nd edn, chapter 5.1.6, John Wiley & Sons, Inc, New York.

[5] Chai, H., Babcock, C.D. and Knauss, W.G. (1981) One-dimensional modeling of failure in laminated plates by delamination buckling. *Int. J. Solids Struct.*, **17**, 1069–1083.

[6] Chai, H. and Babcock, C.D. (1985) Two-dimensional modeling of compressive failure in delaminated laminates. *J. Compos. Mater.*, **19**, 67–98.

[7] Kassapoglou, C. and Hammer J. (1989) Design and analysis of composite structures with manufacturing flaws. Proceedings of 45th AHS Forum and Technology Display, Boston MA, May 1989, pp. 1075–1082. Also in (1990) *J. Am. Helicopter Soc.*, **35**, 46–52.

[8] Kassapoglou C *Design and Analysis of Composite Structures*, 2nd edn, chapter 5.4, John Wiley & Sons, Inc, New York, 2013.

[9] Cairns, D.S. (1987) Impact and post-impact response of graphite/epoxy and kevlar/epoxy structures. PhD thesis. Department Aeronautics and Astronautics, Massachusetts Institute of Technology, Appendix B.

[10] Hellan, K. (1983) *Introduction to Fracture Mechanics*, Chapter 3, McGraw-Hill.

[11] O'Brien, T.K. (1980) Characterization of delamination onset and growth in a composite laminate, in *Damage in Composite Materials*, ASTM STP 775, American Society for Testing and Materials, pp. 140–167.

[12] Kassapoglou, C. (2013) *Design and Analysis of Composite Structures*, 2nd edn, chapter 8.2, John Wiley & Sons, Inc, New York.

[13] O'Brien, T.K. (1991) Local Delamination in Laminates with Angle Ply Matrix Cracks: Part II Delamination Fracture Analysis and Fatigue Characterization. NASA Langley Research Center TM 104076.

[14] Radcliffe, J.G. and Reeder, J.R. (2011) Sizing a single cantilever beam specimen for characterizing facesheet – core debonding in sandwich structure. *J. Compos. Mater.*, **45**, 2669–2684.

[15] ASTM (2013) Standard D5528-13. Standard Test Method for Mode I Interlaminar Fracture Toughness of Unidirectional Fiber-Reinforced Polymer Matrix Composites, ASTM.

[16] Williams, J.G. (1989) End corrections for orthotropic DCB specimens. *Compos. Sci. Technol.*, **35**, 367–376.

[17] O'Brien, T.K., Johnston, W.M. and Tolland, G.J. (2010) Mode II Interlaminar Fracture Toughness and Fatigue Characterization of a Graphite Epoxy Composite Material. NASA TM–2010-216838.

[18] Wang, Y. and Williams, J.G. (1992) Corrections for mode II fracture toughness specimens of composites materials. *Compos. Sci. Technol.*, **43**, 251–256.

[19] Sridharan, S. (ed) (2008) *Delamination Behaviour of Composites*, Woodhead Publishing, Cambridge.

[20] Gillespie, J.W., Jr, Carlsson, L.A., Pipes, R.B. *et al.* (1986) Delamination Growth in Composite Materials. NASA Langley Research Center CR 178066.

[21] Sheinman, I. and Kardomateas, G.A. (1997) Energy release rate and stress intensity factors for delaminated composite laminates. *Int. J. Solids Struct.*, **34**, 451–459.

[22] Williams, J.G. (1988) On the calculation of energy release rates for cracked laminates. *Int. J. Fract.*, **36**, 101–119.

[23] Schapery, R.A. and Davidson, B.D. (1990) Prediction of energy release rate for mixed-mode delamination using classical plate theory. *Appl. Mech. Rev.*, **43**, S281–S287.

[24] Zou, Z., Reid, S.R., Li, S. and Soden, P.D. (2012) General expressions for energy-release rates for delamination in composite laminates. *Proc. R. Soc. A*, **458**, 645–667.

[25] Krüger, R. (2004) Virtual crack closure technique: history, approach, and applications. *Appl. Mech. Rev.*, **57**, 109–143.

[26] Carlsson, L.A. and Kardomateas, G.A. (2011) *Structural and Failure Mechanics of Sandwich Composites*, Springer.

[27] Kassapoglou, C. (2013) *Design and Analysis of Composite Structures*, 2nd edn, chapter 9.2.2, John Wiley & Sons, Inc, New York.

[28] Kassapoglou, C. and Lagacé, P.A. (1986) An efficient method for the calculation of interlaminar stresses in composite materials. *J. Appl. Mech.*, **53**, 744–750.

[29] Brewer, J.C. and Lagacé, P.A. (1988) Quadratic stress criterion for initiation of delamination. *J. Compos. Mater.*, **22**, 1141–1155.

[30] Camanho, P.P., Dávila, C.G. and De Moura, M.F. (2003) Numerical simulation of mixed-mode progressive delamination in composite materials. *J. Compos. Mater.*, **37**, 1415–1438.

[31] Camanho, P.P. and Dávila, C.G. (2002) Mixed-mode Decohesion Finite Elements for the Simulation of Delamination in Composite Materials. NASA Technical Paper 211737.

[32] Dávila, C.G., Rose, C.A. and Camanho, P.P. (2009) A procedure for superposing linear cohesive laws to represent multiple damage mechanisms in the fracture of composites. *Int. J. Fract.*, **158**, 211–223.

[33] Camanho, P.P., Dávila, C.G., Pinho, S.T. and Remmers, J.J.C. (eds) (2008) *Mechanical Response of Composites*, Chapter 4, Springer.

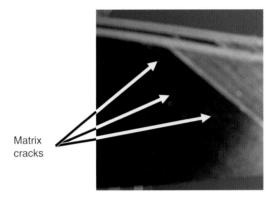

Matrix
cracks

Figure 3.9 Matrix cracks along −45° plies in carbon/epoxy specimen

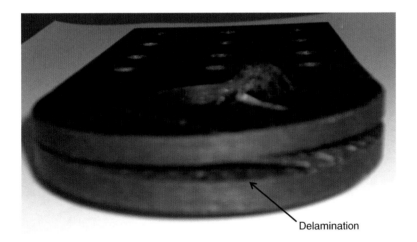

Delamination

Figure 4.1 Delamination in a composite lug after transverse loading

Modeling the Effect of Damage in Composite Structures: Simplified Approaches, First Edition. Christos Kassapoglou.
© 2015 John Wiley & Sons, Ltd. Published 2015 by John Wiley & Sons, Ltd.

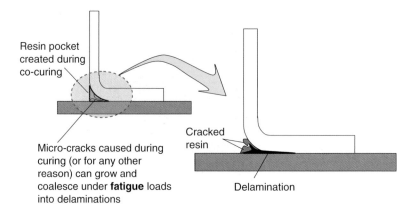

Figure 4.3   Cracks in resin pockets evolving to delamination

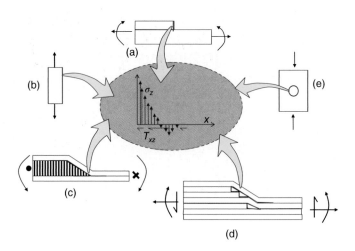

Figure 4.4   Structural details where local inter-laminar stresses may lead to delaminations

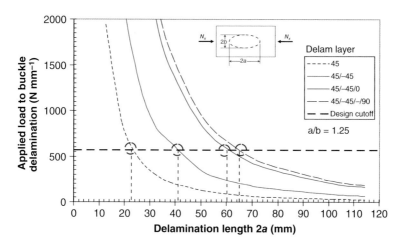

Figure 4.13 Load to cause buckling of delaminating layer in [45/−45/0/90]s laminate

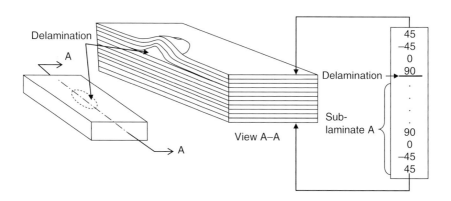

Figure 4.14 Laminate with embedded elliptical delamination

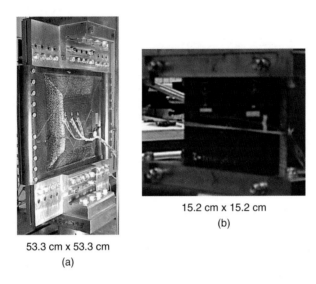

15.2 cm x 15.2 cm

(b)

53.3 cm x 53.3 cm

(a)

Figure 5.7   Sandwich specimens tested under compression

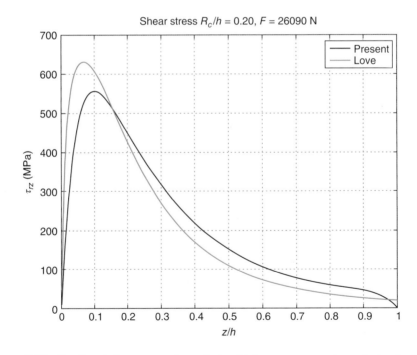

Figure 5.18   Predicted shear stress through the thickness compared to solution by Love

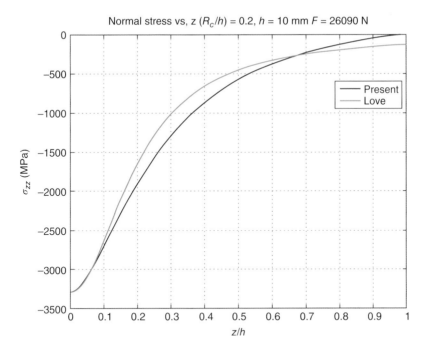

Figure 5.19 Predicted normal stress through the thickness compared to the solution by Love

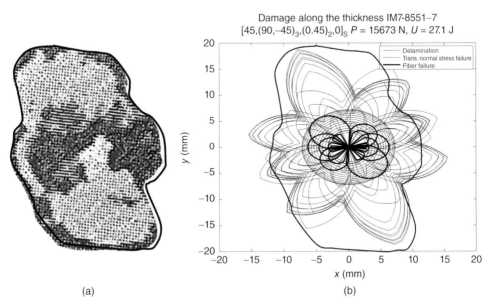

(a)   (b)

Figure 5.27 Ultrasonic scan of damaged region (a) compared to analytical predictions (b) for a [45/(90,−45)$_3$/(0,45)$_2$/0]s IM7/8551-7 laminate with 27.1 J impact

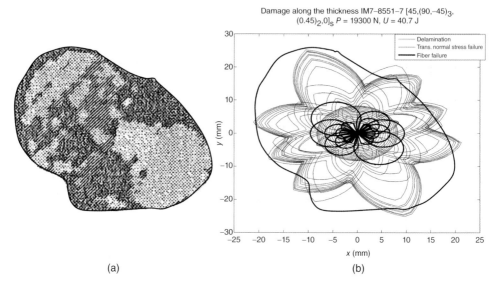

(a)

(b)

Figure 5.28 Ultrasonic scan of damage region (a) compared to analytical predictions (b) for a $[45/(90,-45)_3/(0,45)_2/0]$s IM7/8551-7 laminate with 40.7 J impact

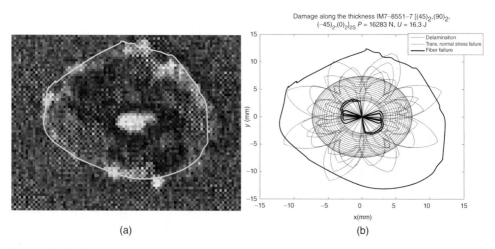

(a)

(b)

Figure 5.29 Ultrasonic scan of damage region (a) compared to analytical predictions (b) for a $[45_2/90_2/-45_2/0_2]_{2s}$ IM7/8551-7 laminate with 16.3 J impact

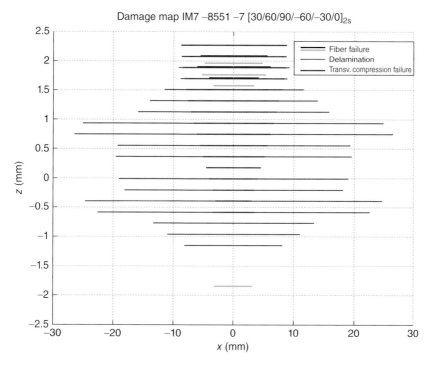

Figure 5.37 Extent and type of damage at $\theta = 0$ for a $[30/60/90/-60/-30/0]_{2}$s IM7/8551-7 laminate

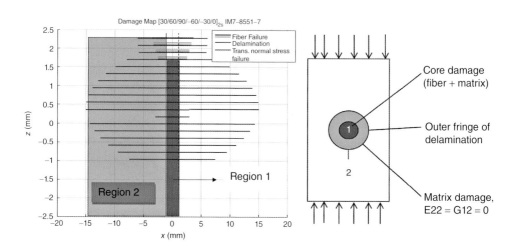

Figure 5.38 Division of damaged laminate into regions

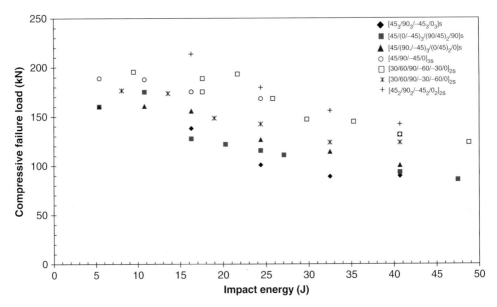

Figure 5.40 Compression after impact (CAI) failure loads for quasi-isotropic laminates (*Source*: From Dost *et al.* [12].)

Figure 6.22 Unidirectional coupon used in static and fatigue tests

5

Impact

Impact damage (see Figure 5.1) combines all the individual forms of damage discussed in previous chapters: holes, cracks, delaminations and fibre breakage. The hope is, therefore, that once the individual types of damage are understood, it would be relatively straightforward to combine the best models of the previous chapters to create a reliable model for impact damage in composites. Unfortunately, nothing is further from the truth. The individual models, as discussed already, are not accurate enough or general enough. In addition, the interaction of the different damage types is not covered in any of the previous chapters, but it turns out it is critical in understanding and modelling impact damage. More elaborate models are necessary for the individual damage types which may then form the stepping stones for a reliable model for impact damage. Until such models are available and readily usable, approximate models for impact damage will form the basis of a design environment.

The impacts considered in this chapter will focus on such combinations of impactor mass and energy, which are typically referred to as *low speed–high mass impacts*. Unlike higher energy impacts where significant amounts of damage are readily visible on the impacted surface, low-speed impact damage shows (almost) no evidence of damage on the surface. Inside the laminate, however, below the impact site, significant damage may be present in the form of matrix cracks, delaminations and broken fibres (Figure 5.1). This damage reduces the compression and shear strength by as much as 60% for first-generation thermoset materials (30–45% for second generation) and becomes an important consideration during the design process.

5.1 Sources of Impact and General Implications for Design

Impact may occur during manufacturing. Tools may be dropped on a part being manufactured, or the part itself may be dropped on the floor or on work benches. In addition, inadvertent collisions of parts with laboratory equipment also can occur.

During service, impact damage may result from various sources: Tool-drops during maintenance, foot traffic, luggage or other equipment drops, hail damage, runway debris thrown up by tyres, or collisions with terrain or ground equipment are some of the possible sources.

Modeling the Effect of Damage in Composite Structures: Simplified Approaches, First Edition. Christos Kassapoglou.
© 2015 John Wiley & Sons, Ltd. Published 2015 by John Wiley & Sons, Ltd.

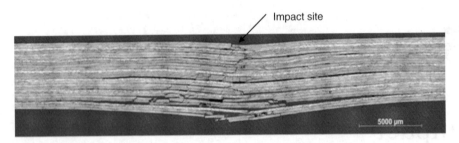

Figure 5.1   Section cut showing the damage created in a quasi-isotropic laminate after 25 J impact

It is obvious that these impacts, whether they are caused during manufacturing or during service, cover a wide range of energies and impactor shapes. Also, the likelihood of occurrence of each type of impact is different. Creating a threat scenario that associates different impact energies with specific probabilities of occurrence is one way of quantifying the threat and determining what condition to design for. Typical impact threats during manufacturing are summarised in Table 5.1. The impact energies shown in Table 5.1 correspond to drops from most likely height.

Typical energies and their likelihood of occurrence for impacts during service are shown in Figure 5.2. It can be seen from the figure that more common threats such as tool-drops, dropped parts and foot traffic correspond to low energies not exceeding 25 J. At the other end, the highest impact energy is around 60 J, which corresponds to impacts with terrain or equipment on the ground.

The threats presented in Table 5.1 and Figure 5.2 may be considered as typical. In addition to these, there are rare, extreme cases that must be taken into account. The drop of an entire toolbox has been proposed as the most extreme case. This corresponds to 135 J.

Table 5.1   Impacts from dropped tools

Tool	Energy (J)	Blunt impact	Sharp impact
Metal ruler	3.0	—	X
Measuring calipers	3.0	—	X
Ratchet wrench	4.1	X	—
Drill bit	5.0	—	X
Soft hammer	6.0	X	—
File	6.0	X	—
Pliers	7.1	X	—
Screw driver	7.1	—	X
Wrench	10.0	X	—
Adjustable wrench	12.1	X	—
Rivet gun	24.9	X	—
Power equipment	24.9	X	—

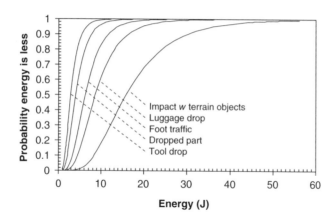

Figure 5.2 Energy levels and probabilities of occurrence for various impact threats

Depending on the type of threat and energy level, the damage created may be easily detectable or not. Detection depends on the inspection method selected. More advanced methods, such as ultrasonic inspection or X-rays, can detect smaller damage sizes (see Figure 4.6). Simpler methods such as visual inspection are less reliable and detect larger damage sizes. In addition, sharp impacts (see Table 5.1) tend to puncture the surface, which makes detection easier, while blunt impacts tend to leave no trace on the surface and require more elaborate inspection methods for detection.

Given this range of damage threats and types and the wide variety of inspection methods (see Figure 4.6), some consistency in relating resulting damage to load-carrying ability is necessary. Once the inspection method has been selected, any type of damage that is not detectable by that specific method may be present in the structure and there is no way of knowing about it. From the perspective of the user, the structure is 'pristine' and thus must be able to carry the ultimate load. This means that structure with damage up to the 'threshold of detectability' (TOD) of the selected inspection method is treated as 'pristine' structure and must not fail under ultimate load.

On the other hand, damage easily detectable by the selected inspection method is readily known to the user upon inspection. This could be impact damage with a sharp impactor (see Table 5.1) which is visually detectable. Then, the factor of safety of 1.5 between limit and ultimate load, which, among other reasons, is meant to cover uncertainties related to unknown damage present, need no longer be applied. Therefore, a structure with detectable damage must meet the limit load. Note that this does not include extensive, discrete source damage such as the one that may be caused by bird or lightning strike, which is immediately known to the pilot and, as such, requires less than limit load capability, the so-called safe-return-to-base loads.

It is common practice to use visual inspection as the method for routine inspections. Then, the TOD, mentioned earlier, becomes barely visible impact damage (BVID). This means that composite structure with BVID must meet the ultimate load. If the damage is visible, the structure must meet limit load.

A common problem with BVID is that it is a function of the person doing the inspection, the lighting conditions and the paint (or the lack of it) used on the impacted surface. In an attempt to make the definition of BVID more objective, it has been proposed that BVID be defined as the damage with indentation depth of 1 mm seen from approximately 1 m away. It should be pointed out that the dent depth relaxes with time. For this reason, BVID is typically defined as damage leaving a dent of 1 mm depth after three days so that most of the relaxation has already taken place.

The effect of impact damage on compression strength is compared with other types of damage in Figure 5.3.

As can be seen from Figure 5.3, impact damage is the most critical type of damage among the ones that are not visually detectable: void content, porosity, delamination and impact damage. Holes are the most critical among visually detectable damage, such as flawed holes and open holes. This suggests that, if one uses BVID to design for ultimate load and a certain diameter hole for limit load, most types of damage caused during manufacturing or service are accounted for.

One might be tempted to not use BVID as a design condition for some composite parts that are protected from impact. For example, parts enclosed in other parts or not exposed to typical threats such as hail damage or runway debris could be considered exempt from the BVID design requirement. However, this is not the correct approach. It is impossible to guarantee that during manufacturing and before the parts in question are enclosed by other parts, they will not be impacted inadvertently. Or, during overhaul maintenance when part of the structure is disassembled, these parts are exposed to impact threats again. For this reason, BVID must be considered in the design of every composite part. In addition, because of the randomness of impact events, it is impossible to guarantee that BVID will not occur at the most highly loaded location of the part being designed. As a result, the design must account for BVID at the most highly loaded location even if this is very unlikely.

In addition to the difficulties in accurately defining it, as mentioned above, the definition of BVID causes problems in designing thick composite structures. As the

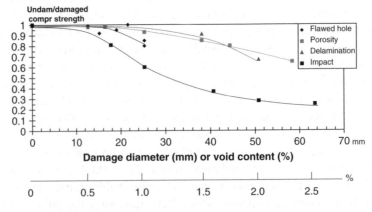

Figure 5.3   Effect of damage on compression strength [1] (*Source*: Kassapoglou [1].)

thickness of the structure increases, the energy required to cause BVID, as defined by the 1-mm dent depth, increases. For a sufficiently thick structure, the energy required may be 'too high', exceeding any energy level expected during the life of the structure. In such a case, using BVID as an ultimate load requirement for thick structure is overly conservative. For a more representative design, the highest value between the cut-off energy of 135 J mentioned earlier and the highest energy from a threat scenario representing realistic threats over the life of the structure should be used.

5.2 Damage Resistance Versus Damage Tolerance

There are two distinct yet inter-related issues associated with impact. The first is the extent and type of damage created, and the second is the strength of the structure in the presence of this damage.

It is important to recognise that the extent of damage is not uniquely related to the residual strength. The extent of damage is, usually, measured as an overall size enveloping delaminations and fibre breakage. This is, typically, what is obtained from ultrasonic inspection. Even if the exact through-the-thickness location of each delamination is known from inspection, the extent of damage may be a misleading predictor of residual strength. An example is shown in Figure 5.4.

Multiple small delaminations are created after impact in Figure 5.4a. If a compression load is now applied, all or most of the resulting sub-laminates will buckle early because they have a low bending stiffness. Final failure will follow soon after buckling. The situation is different in Figure 5.4b. Even though the delamination is larger, the resulting sub-laminates are thick and have very a high bending stiffness compared to the multiple sub-laminates of Figure 5.4a. Therefore, the sub-laminates in Figure 5.4b will buckle at a much higher load and the final failure load for the laminate in Figure 5.4b will be higher.

The example of Figure 5.4 is only one of the many situations where larger damage extent does not necessarily imply lower compression after impact (CAI). It is therefore, important to differentiate between the mechanisms associated with the amount of

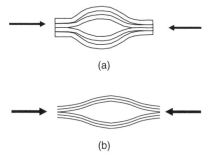

(a)

(b)

Figure 5.4 Special cases of damage extent caused by impact. (a) Multiple small delaminations and (b) single larger delamination

damage created and the mechanisms that lead to final failure once a certain amount of damage is present. The two situations are identified as *damage resistance* and *damage tolerance*, respectively. Damage resistance measures the ability to contain the extent and type of damage given a specific threat. In the case of impact, the specific threat will be impact with an impactor of a specific geometry and at a specified energy level. Damage tolerance measures the ability to reach a certain load, without failure, given a specific amount of damage.

As already suggested, good damage resistance does not always imply good damage tolerance. However, in the absence of other means to quantify damage tolerance (accurate analysis or test results), coming up with a design that has good damage resistance is a good first step towards good damage tolerance. It is within this context that toughened matrix materials were developed. They result in excellent damage resistance, which in some cases translates to good damage tolerance. Thermoplastic materials are one example. The tougher thermoplastic matrix keeps large delaminations from forming during impact.

Taken to an extreme, increased toughness may give opposite results [2]. For example, consider the case of a thermoplastic laminate designed with BVID. Because of the fact that the matrix is tougher, higher energy is needed to create the indentation corresponding to BVID. This means that there will be more broken fibres in a thermoplastic laminate with BVID than a thermoset laminate with its BVID. As a result, there are fewer intact fibres to carry the compression load, and the CAI strength may be lower.

For simplicity, and in order to standardise experiments so that results from different laminates can be comparable, the impactor is usually assumed to be made of steel with a spherical tip of diameter 0.8 cm to 2.5 cm. Also, standard test specimen configurations have been defined [3], but these are of limited use in design because their dimensions, boundary conditions and thickness do not correspond to typical structural composite parts.

The complete solution to determining the effects of impact damage would then consist, first, of the damage resistance step where the geometry of impactor and impact energy are used as inputs to determine the type, extent and through-the-thickness location of damage for a given laminate. This is then followed by the damage tolerance step for determination of the residual strength under a given loading.

The complexity of this problem makes it necessary to use modelling approaches that are computationally very intensive given the power of today's computers. These methods are thus not very useful for design and optimization when a large number of different laminates might be compared for each location in the structure in order to arrive at the optimum performer. As a result, simpler and more efficient, but also less accurate, methods have been developed over the years to assist the designer/analyst in selecting the best candidates and minimising the subsequent test effort to verify the design. In subsequent sections, some of these simpler approaches will be presented in order of increasing complexity. These are (i) modelling impact damage as a hole, (ii) modelling impact damage as a delamination, (iii) modelling impact damage as an area of reduced stiffness and (iv) incorporating damage as part of the continuum model used to analyse the structure.

5.3   Modelling Impact Damage as a Hole

In the extreme case where the impact energy is high enough so that the impactor completely penetrates the laminate, the damage will consist of a hole of an irregular shape with additional limited damage around its circumference. One can then model the resulting damage as a hole of an equivalent size. This approach has been used successfully to model ballistic damage [4]. Thus, modelling impact damage as a hole [5, 6] is, in a sense, an extension of this approach for ballistic damage to lower impact energies.

The idea is that there will always be a hole size that will result in the same residual strength as a given impact damage. Then, analysis of holes, which is easier than that of impact damage, can be used, following, for example, the approaches discussed in Chapter 2. The situation is shown in Figure 5.5.

Referring to Figure 5.5, the main question is how to determine the size of the equivalent hole. In general, the equivalent hole will be elliptical with major and minor axes $2a$ and $2b$, respectively. These must be somehow related to the overall dimensions of the impact damage $2A$ and $2B$, where again the impact damage as measured, for example by ultrasonic scan, is also modelled as elliptical.

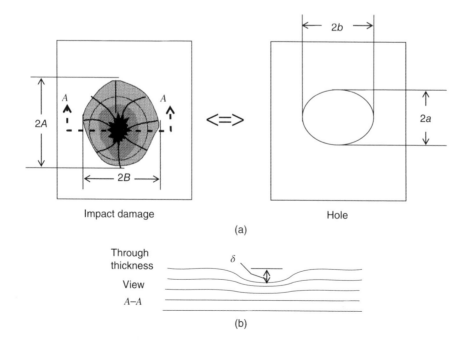

Figure 5.5   (a,b) Modelling impact damage as a hole

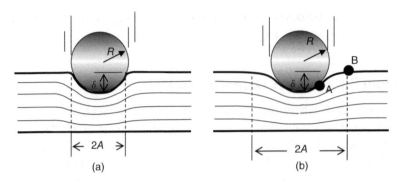

Figure 5.6 (a,b) Possible shapes of impacted surface during impact. Indentation is the same in both cases

For the case of CAI, one approach, proposed by Puhui *et al.* [6], sets the dimension 2*b* perpendicular to the applied compressive load, equal to 2*B*. The dimension 2*A* is then determined as a function of the indentation δ and the impactor radius *R*. For example, if the impactor is assumed perfectly rigid compared to the laminate, the shape of the indentation can be assumed to follow the shape of the impactor as shown in Figure 5.6.

In the extreme case in Figure 5.6a, the top surface of the impacted laminate is in complete contact with the impactor over the entire surface portion of the impactor corresponding to the indentation δ. This is only approximated by very soft laminates. Usually, only a portion of the top surface is in contact with the impactor, as shown in Figure 5.6b.

The dimension 2*A* is then related to the region under large deformations in the impacted laminate. For the case of Figure 5.6a, this dimension is approximated as

$$2A = 2\sqrt{R^2 - (R - \delta)^2} = 2\sqrt{2R\delta - \delta^2} \tag{5.1}$$

For the case to the right of Figure 5.6, additional information is needed about the shape of the curve AB, which connects the tangency point A between the impactor and the laminate, and the point B, which defines the beginning of the flat region on the top surface of the impacted laminate. Different curve shapes can be assumed, or experimental data can be used to help define this curve.

With 2*A* and 2*B* defined, the impacted laminate under compression can be analysed as a laminate with an elliptical hole with the dimension 2*A* parallel to the loading direction. If 2*A* and 2*B* are approximately equal, the equivalent hole is circular and analysis methods from Chapter 2 can be used directly. If 2*A* is significantly different than 2*B*, the Whitney–Nuismer approach in Chapter 2 can still be used but the stress distribution near the hole edge must now be determined via an anisotropic elasticity solution [7, 8].

For the case of sandwich structures, this equivalence of a hole to impact damage can be defined more precisely for a certain class of laminates and a specific impact level,

BVID. If the facesheet lay-up is close to quasi-isotropic, where the facesheet stiff-ness in any direction is within 20% of the corresponding stiffness of a quasi-isotropic laminate, BVID can be approximated with a 6.35-mm-diameter hole.

Tests were performed on sandwich panels of two different sizes, 15.2 cm × 15.2 cm and 53.3 cm × 53.3 cm with 6.35 mm holes or BVID. The facesheets included quasi-isotropic laminates as well as laminates that were approximately 20% stiffer or 20% softer. The test specimens are shown in Figure 5.7. Note that the large-sized specimen had a ramp-down all around transitioning from sandwich to monolithic laminate, while the small-sized specimens had constant thickness with potting compound at the loaded ends.

Two important conclusions were drawn from the tests:

1. The smaller size specimens with BVID failed, on average, at 67% of the undam-aged static strength, while the bigger size specimens had an average failure of 76% of the undamaged static strength. This difference is outside experimental scatter. This means that there is a size and boundary conditions effect. Smaller specimens with damage fail at lower loads. There are several reasons for this. Larger specimens absorb more impact energy by deflecting out of plane instead of having damage created. Smaller specimens may have some finite width effects. The boundaries where the specimen is held, which locally stiffen the structure, are closer to the impact site for smaller specimens.
2. The bigger size specimens with BVID and the OHC (open hole compression) spec-imens of the same size with 6.35 diameter holes have the same knock-down in strength. Statistical analysis of the tests showed that the two sets of data, BVID and OHC, can be treated as the same pool of data with the same mean strength at 76%

53.3 cm x 53.3 cm   15.2 cm x 15.2 cm

(a)   (b)

Figure 5.7 Sandwich specimens tested under compression (*See insert for colour represen-tation of this figure.*)

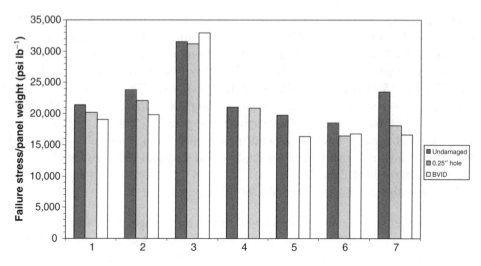

Figure 5.8   (a,b) Compression strength of 53 cm × 53 cm sandwich specimens with BVID or 6.35 mm hole compared to undamaged strength for various lay-ups and cores (English units are used.)

of the undamaged static strength. Therefore, a 6.35-mm hole can be considered as having the same residual strength as BVID for these sandwich configurations. A summary of the results for facesheet lay-ups and core materials and densities is shown in Figure 5.8. These results cover facesheet thicknesses up to ∼1 mm.

A similar situation was observed for shear after impact (SAI). Test results from 53.3 cm × 53.3 cm specimens with a 6.35-mm hole were statistically indistinguishable from the corresponding results of specimens with BVID. Therefore, analysis methods for open holes under shear loads can be used to replace analysis methods for SAI. Again, this conclusion is valid for nearly quasi-isotropic facesheets with thickness up to 0.8 mm. The test specimens and a subset of test results for two different combinations of core density and facesheet lay-up are shown in Figure 5.9.

5.4   Modelling Impact Damage as a Delamination

In this case, it is recognised that the damaged region is still capable of carrying load. As a first approximation, instead of attempting to model all delaminations created during impact, a single delamination of equivalent size is placed at a specific ply interface [9, 10]. The situation is shown schematically in Figure 5.10. The load at which one of the sub-laminates on the right of Figure 5.10 buckles must be equal to the CAI load on Figure 5.10a. Two problems arise: What is the equivalent delamination size, and where through the thickness that delamination should be located.

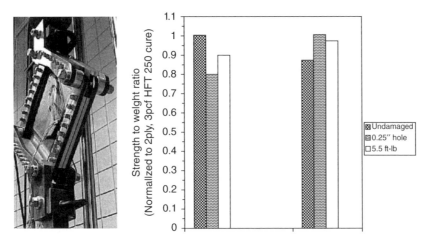

Figure 5.9   (a,b) Shear strength of $53\,cm \times 53\,cm$ sandwich specimens with BVID or 6.35 mm hole compared to undamaged strength for various lay-ups and cores (English units are used.)

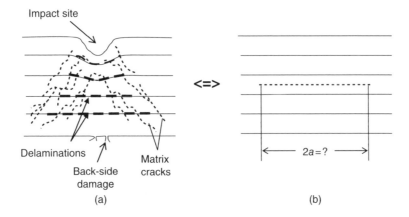

Figure 5.10   (a,b) Simulating impact damage with a single delamination

The determination of the equivalent delamination size and location depends strongly on the lay-up and thickness. Extensive tests are necessary to establish usable values that give good approximation of the CAI load for a wide enough range of stacking sequences.

For the specific case of sandwich laminates with up to 1.3 mm facesheet thickness, tests have shown that the size of the delamination caused by impact increases with distance from the point of impact. The largest delamination usually is found between the ply next to the core and its neighbour. In some cases, where the adhesive

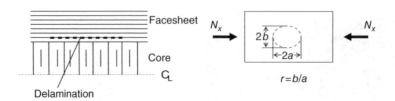

Figure 5.11   Equivalent delamination representing BVID in sandwich structure

connecting the core to the facesheet is not strong enough, larger disbonds between core and facesheet may be present; here the adhesive bond is assumed strong enough so the largest delamination is located at the first ply interface away from the core as shown in Figure 5.11. If there are only two plies per facesheet, then impact up to BVID damages the core below the impact site, causing an equivalent disbond between the core and facesheet. So in cases where facesheets with two plies are used, the equivalent delamination is a disbond between the facesheet and core.

As already implied, the delamination representing impact damage is assumed elliptical in shape. Its aspect ratio b/a is assumed known. Typically, it is the same aspect ratio as the ellipse circumscribing BVID in the structure being simulated. This aspect ratio can be obtained from tests on the same laminate or similar laminates. This leaves $2a$ as the unknown parameter in this model.

The value of the delamination length $2a$ is obtained by relating buckling of the delaminating layer in the facesheet to other failure modes in the sandwich when there is no damage or delamination. If there is no damage, other sandwich failure modes, such as facesheet strength, wrinkling, dimpling or panel buckling [11], must be evaluated to determine the critical failure mode. Then, a delamination of aspect ratio b/a is placed at the critical location of Figure 5.11. Its length $2a$ is varied until buckling of the elliptical delaminating layer above this delamination occurs at an overall applied load which equals the undamaged failure load of the sandwich in compression. This size corresponds to the situation where failure due to delamination buckling coincides with failure of the undamaged panel. To account for the presence of BVID, this size is now increased by a factor 1.22. This factor was experimentally determined over a wide variety of sandwich laminates [9]. Note that, if the damage is not BVID, a different factor must be used. Thus, CAI in a sandwich with up to 12 plies per facesheet and lay-up with stiffness in any direction within 20% of that for a quasi-isotropic layup is approximated as follows:

1. Determine the failure load of the sandwich under compression if there is no damage present.
2. Determine or decide on the aspect ratio b/a of the damage caused by barely visible impact (see Figure 5.11 for definition of b/a).
3. Locate a delamination of aspect ratio b/a at the interface between the ply next to the core and its neighbour (see Figure 5.11).

4. Determine the size $2a$ of the delamination in step 3 such that the buckling load of the delaminating layer consisting of the plies above the delamination in Figure 5.11 coincides with the failure load in step 1.
5. Increase the size $2a$ by 1.22 and determine the CAI strength as the buckling stress of the delaminating layer with the new value of $2a$.

5.5 Impact Damage Modelled as a Region of Reduced Stiffness

The next logical step in improving the approach to model impact damage is to model the damage site as a region of reduced stiffness. The matrix cracks, broken fibres and delaminations of the damaged region change the in-plane stiffness of the laminate locally. In general, the stiffness of the damaged region is not constant. It increases from a low value or zero value (if there is puncture) at the centre of the damage region to the value of the undamaged laminate at the boundary of the damaged region. This stiffness variation is not necessarily the same along different radial lines emanating from the centre of the damaged region. As a first approximation, to be relaxed later, the stiffness reduction of the damaged region will be assumed constant. This means that, if the undamaged region is described by laminate stiffnesses $E_{11}^L, E_{22}^L, G_{12}^L, v_{12}^L$, given by Equation 2.4a–d, the damaged region will be described by stiffnesses $rE_{11}^L, rE_{22}^L, rG_{12}^L, rv_{12}^L$, where r is the constant stiffness reduction caused by impact damage, also referred to as *modulus retention ratio* [12]. The situation is shown in Figure 5.12. The ratio r depends on the energy level. For high impact energies, it is closer to zero, while for low energies it will approach 1.

The problem has now been reduced to determining the stresses in an orthotropic plate with an orthotropic elliptical inclusion with stiffnesses a fraction r of the stiffnesses of the material surrounding the inclusion. This has been solved by Lekhnitskii [13] using complex variables. If $r < 1$, the stresses are maximised at the boundary

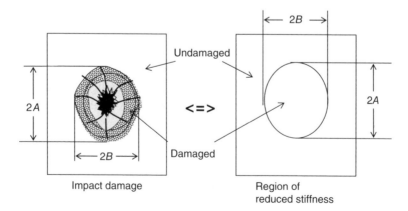

Figure 5.12 Region of impact damage approximated as a region of reduced stiffness

of the inclusion. For the special case where the inclusion is nearly circular, the stress concentration factor (SCF) describing the stresses at the boundary of the damaged region can be shown to be

$$
\text{SCF} = 1 - (1 - \lambda) \frac{
\begin{aligned}
&1 + (\lambda + (1 - \lambda)v_{12}^2(E_{22}/E_{11}))\sqrt{2(\sqrt{E_{11}/E_{22}} - v_{12}) + (E_{11}/G_{12})} \\
&+ ((E_{11}/G_{12}) - v_{12})\sqrt{E_{22}/E_{11}}
\end{aligned}
}{
\begin{aligned}
&1 + \lambda[\lambda + (1 + \sqrt{E_{22}/E_{11}})\sqrt{2(\sqrt{E_{11}/E_{22}} - v_{12}) + (E_{11}/G_{12})}] \\
&+ ((E_{11}/G_{12}) - 2\lambda v_{12})\sqrt{E_{22}/E_{11}} - (1 - \lambda)^2 v_{12}^2(E_{22}/E_{11})
\end{aligned}
}
$$

$$(5.2)$$

where E_{11}, E_{22}, G_{12}, v_{12} are the in-plane laminate stiffness quantities for the undamaged laminates, and $\lambda = 1/r$ is the ratio of the undamaged to damaged stiffness with $\lambda > 1$.

The SCF given by Equation 5.2 is for an infinite plate. If the damaged region makes up a significant portion of the impacted plate, a finite width correction factor must be used along with SCF. As a first approximation, Equation 2.5 can be used.

It is interesting to note that Equation 5.2 reproduces known results for some limiting cases. For example, if the damaged region is a hole, $r = 0$ and $\lambda = \infty$. Substituting in Equation 5.2 gives

$$
\text{SCF} = 1 + \sqrt{2\left(\sqrt{\frac{E_{11}}{E_{22}}} - v_{12}\right) + \frac{E_{11}}{G_{12}}} \quad \text{open hole} \tag{5.3}
$$

which is the same as Equation 2.1 evaluated at $\theta = 90°$.

Also, for the case where there is no impact damage, that is, $r = 1$, substituting in Equation 5.2 gives SCF = 1 as expected.

Once the SCF has been determined, the CAI strength can be obtained from

$$
\sigma_{\text{CAI}} = \frac{\sigma_c^u}{\text{SCF}} \tag{5.4}
$$

where σ_c^u is the ultimate compression strength of the laminate without damage.

The modulus retention ratio r is still unknown. As was already mentioned, it depends on the impact energy. Usually, some additional analysis (see Section 5.7.2.2 below) and/or test results are needed to determine it. A qualitative discussion that gives some insight to its value is given here.

In general, the stiffness in the damaged region can be as low as 0 at the centre or a region around it and increases to 1 at the edge of the damaged region. If this variation of stiffness is assumed to be linear, possible representations of the stiffness in the damaged region are shown in Figure 5.13. If the impact energy is sufficiently high, the impactor creates a hole at the centre of the impact region. The stiffness is zero there. It starts from zero at the edge of the hole created during impact and goes to the far-field stiffness E_{ff} of the undamaged region, as shown in Figure 5.13a. If the

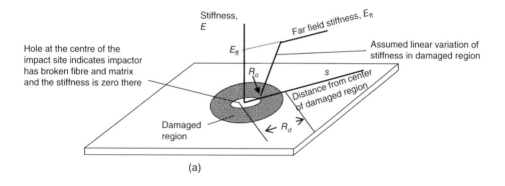

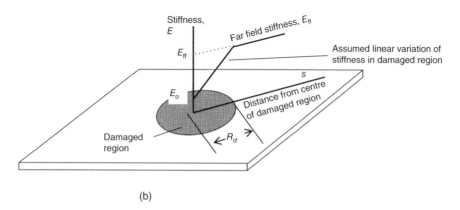

Figure 5.13 Possible representations of stiffness in the damage region. (a) High impact energy (puncture in the middle of the damage region) and (b) low impact energy

impact energy is relatively low, the stiffness at the centre of the impact site is greater than zero but less than E_{ff}. This is shown in Figure 5.13b.

If the starting value of the stiffness E_o in the damaged region is known, then a linear distribution between that value and E_{ff} can be used to approximate the average stiffness along a radial line as

$$E_{dam} = \frac{1}{R_d} \int_0^{R_d} \left(E_o + (E_{ff} - E_o) \frac{s}{R_d} \right) ds \qquad (5.5)$$

Note that, if there is a hole at the centre of the damaged region, the lower limit of the integral of Equation 5.5 must be replaced by the radial distance R_o which defines the end of the hole as shown in Figure 5.13a.

A more quantitative evaluation can be obtained if the SCF is evaluated for different r values (modulus retention ratio) for various lay-ups. This is shown in Figure 5.14. The lay-ups used in Figure 5.14 range from highly orthotropic (all 0 unidirectional tape) to matrix-dominated (all 45 fabric material).

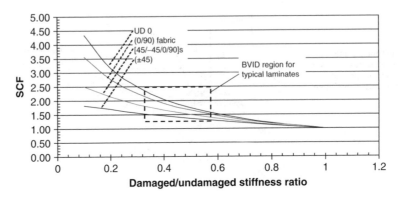

Figure 5.14 Stress concentration factor at the edge of circular elastic inclusion as a function of stiffness ratio r and stacking sequence

Several interesting conclusions can be drawn from Figure 5.14. The first is that the all-fabric (±45) laminate has the lowest SCF, which for BVID is in the range of 1.2–1.5. This characteristic of all 45/−45 laminates to have very low SCFs in the presence of notches was also observed in Table 2.1, where the experimentally measured SCF for such laminates with an open hole is less than 1.1. Based on this observation, increasing the number of 45 plies in a laminate reduces the SCF caused by impact damage (or open hole). It should be emphasised, however, that a laminate consisting solely of 45/−45 plies does not have the highest CAI for a given thickness and impact energy. Such a laminate has the lowest SCF but at the same time has very low undamaged compression strength. This means that both the numerator and the denominator of Equation 5.4 decrease. In general, increasing the percentage of all 45° plies in a laminate decreases the undamaged strength faster than the SCF in the presence of impact damage. The optimum lay-up from a CAI strength perspective is one that combines 0° and 45° plies. The 0° plies increase the undamaged strength and the 45° plies decrease the SCF after impact. This effect was examined in some detail in the application in Section 2.7 where combinations of 0° and 45° plies that combine high strength and low SCF were determined. Of course, in addition to these two orientations, other orientations are usually necessary to satisfy other load requirements and design guidelines such as, for example, the 10% rule.

The second conclusion drawn from Figure 5.14 is that, if one excludes the all 0 laminates, which is very rarely used in practice, the stiffness of the remaining laminates covers the range of expected results in practice reasonably well. This means that for BVID the SCF would be in the range 1.2–2.2.

The third and final conclusion drawn from Figure 5.14 relates to a difference between damage resistance and damage tolerance. The (0/90) fabric and the (±45) fabric laminates are identical in stacking sequence. The only difference is that one is rotated 45° relative to the other. This means that for a given impact energy, they both have exactly the same type and extent of damage. However, the SCF is significantly different as can be seen by comparing the second curve from the top with the lowest curve in Figure 5.14. This is consistent with the fact that the (0/90) laminate has fibres aligned with the direction of loading while the (±45) laminate does not.

It is important to keep in mind that the results from Equations 5.3 and 5.4 and Figure 5.14 are approximate. Equation 5.3 holds for circular damage. If the damage contour departs significantly from a circular shape, the results presented are not expected to hold. This would be the case of an all 0 unidirectional laminate (top curve in Figure 5.14), where tests have shown that the damage is elongated and aligned with the fibre direction including longitudinal splits parallel to the fibres.

Another aspect worth mentioning in relation to Equation 5.3 is that it does not distinguish between laminates with the same plies but different stacking sequences. Only the in-plane laminate stiffnesses E_{11}, E_{22}, G_{12} and v_{12} are present in Equation 5.3, which do not change if the stacking sequence is reshuffled. This can be an important limitation, as will be discussed in Section 5.7.2.3 and Figure 5.40.

Finally, this discussion is based on strength reduction and neglects completely buckling of the delaminated sub-laminates. If the delaminations are large enough, which happens for moderate to high impact energies per unit laminate thickness, delamination buckling may occur before failure as a result of the stress concentration effect of the damage region and may precipitate final failure. In such a case, the strength analysis must be complemented with a buckling analysis similar to the one presented in Section 4.3.2.

5.6 Application: Comparison of the Predictions of the Simpler Models with Test Results

To get a feel of how the three simpler models, as a hole, as an equivalent delamination and as a region of reduced stiffness, perform in predicting impact damage, they are compared with test results in this section.

Three different facesheet layups:

A: [(±45)/(0/90)]
B: [(±45)/(0/90)/(±45)]
C: [(±45)/(0/90)$_2$/(±45)].

were used to manufacture 15.2 cm × 15.2 cm sandwich laminates with a 2.5-cm core. The specimens were potted at the two opposite ends and were impacted, each at its corresponding BVID level. They were then tested in compression using the fixture shown in Figure 5.7b.

The basic material properties for the plain weave fabric material used were as follows:

$$E_x = 73.0\,\text{GPa}$$
$$E_y = 84.0\,\text{GPa}$$
$$G_{xy} = 5.3\,\text{GPa}$$
$$\nu_{xy} = 0.05$$
$$\text{ply thickness} = 0.1905\,\text{mm}$$

The experimentally measured undamaged failure strengths under compression were as follows:

A: 328.1 MPa
B: 297.5 MPa
C: 291.3 MPa.

5.6.1 Modelling BVID as a Hole

In order to model BVID as a hole, it suffices to assume that instead of BVID a 6.35-mm-diameter hole is present in accordance with Section 5.3. Then Equation 2.11 can be used. In order to apply Equation 2.11, the SCF for an infinite plate with the same lay-up as the facesheet containing a 6.35-mm-diameter hole must be obtained. For this, Equation 2.6 is used. These SCFs are shown in the last column of Table 5.2.

Applying the procedure mentioned briefly in Section 2.6 for the determination of the compressive characteristic distance, the value of $d_o = 2.1\,\text{mm}$ is found. Equation 2.11 is applied to determine the ratio of stress at d_o from the hole edge to the far-field applied stress for the three lay-ups. The corresponding ratio for each of the laminates is as follows:

A: 1.380
B: 1.398
C: 1.380.

Table 5.2   SCF for infinite plates with a hole

Facesheet layup	E_{11} (Pa)	E_{22} (Pa)	G_{12} (Pa)	ν_{12}	k_t^∞
(±45)/(0/90)	53,773,200,000	58,736,880,000	2.1234E + 10	0.307	2.96
(±45)/(0/90)/(±45)	44,604,180,000	47,568,600,000	2.6542E + 10	0.429	2.66
(±45)/(0/90)₂/(±45)	53,773,200,000	58,736,880,000	2.1234E + 10	0.307	2.96

The far-field CAI stress is then obtained as the far-field stress, which makes the stress at d_o equal to the undamaged failure strength given earlier. Thus, for laminate A

$$\frac{\sigma(y = d_o)}{\sigma_{\text{far field}}} = 1.38 \Rightarrow \frac{\sigma_c^u}{\sigma_{\text{CAI}}} = 1.38 \Rightarrow \sigma_{\text{CAI}} = \frac{328.1}{1.38} = 237.7\,\text{MPa}$$

Similarly, approximations to the CAI failure stress can be obtained for the remaining two lay-ups. The results are summarised below:

Facesheet lay-up	σ_{fCAI} (MPa)
$(\pm 45)/(0/90)$	237.7
$(\pm 45)/(0/90)/(\pm 45)$	212.8
$(\pm 45)/(0/90)_2/(\pm 45)$	211.0

5.6.2   Modelling BVID as a Single Delamination

The delaminations are assumed to be circular on the basis of ultrasonic inspection showing a nearly circular damage region. Using Section 5.4, if $2a$ is the delamination length that would cause delamination buckling at the same far-field load as the undamaged failure strength

$$\sigma_c^u = \frac{f(\text{layup})}{(2a)^2}$$

where $f(\text{layup})$ is a function of the lay-up of the delaminating layer and the overall facesheet involving the D (bending stiffness) and α (inverse of ABD) matrices. According to Section 5.4, increasing $2a$ by a factor of 1.22 makes the delamination buckling load equal to the CAI stress:

$$\sigma_{\text{CAI}} = \frac{f(\text{layup})}{(1.22a)^2}$$

Combining the two equations to eliminate the term $f(\text{layup})/(2a)^2$, we get

$$\sigma_{\text{CAI}} = \frac{\sigma_c^u}{1.22^2}$$

Applying this equation gives the CAI predictions as follows:

Facesheet lay-up	Undamaged fail str (MPa)	(MPa) CAI Pred fail str
$(\pm 45)/(0/90)$	328.08	220.43
$(\pm 45)/(0/90)/(\pm 45)$	297.46	199.85
$(\pm 45)/(0/90)_2/(\pm 45)$	291.34	195.74

5.6.3   Modelling BVID as an Elliptical Inclusion of Reduced Stiffness

As mentioned in Section 5.5, in order to use this model, the average stiffness of the damaged region must be known. In the case of the three sandwich laminates under discussion, BVID created a tiny pinhole at the centre of the impact site. This means that the stiffness of the damaged region is zero at the centre because of the pinhole and equal to the undamaged stiffness at its edge. If the variation of stiffness in the damaged region is assumed to follow a straight line, as shown in Figure 5.15, the average stiffness along a radial line is half the undamaged stiffness.

The following properties are calculated for the three lay-ups:

Facesheet lay-up	E_{11}	E_{22}	G_{12}	v_{12}
(±45)/(0/90)	53.77 E9	58.74 E9	21.23 E9	0.307
(±45)/(0/90)/(±45)	44.60 E9	47.57 E9	26.54 E9	0.429
(±45)/(0/90)₂/(±45)	53.77 E9	58.74 E9	21.23 E9	0.307

Then, substituting in Equations 5.2 and 5.4, the CAI strength predictions are obtained as follows:

Facesheet lay-up	Undamaged fail stress (MPa)	SCF	Prediction fail stress (MPa)
(±45)/(0/90)	328.08	1.49339	219.69
(±45)/(0/90)/(±45)	297.46	1.45908	203.87
(±45)/(0/90)₂/(±45)	291.34	1.49339	195.09

5.6.4   Comparisons of Analytical Predictions to Test Results – Sandwich Laminates

The predictions from the three methods, namely modelling as a hole, delamination and region of reduced stiffness, are compared to test results in Table 5.3.

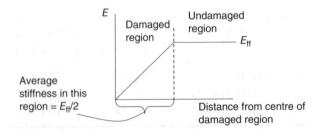

Figure 5.15   Assumed stiffness variation in the damaged region

Table 5.3 CAI strength predictions compared to test results for sandwich with BVID

Facesheet layup	Test failure (MPa)	Damage as hole (MPa)	% difference	Damage as delam (MPa)	% difference	Damage as red stiff (MPa)	% difference
(±45)/(0/90)	254	237.657	−6.4	220.43	−13.2	219.69	−13.5
(±45)/(0/90)/(±45)	201	212.799	5.9	199.85	−0.6	203.87	1.4
(±45)/(0/90)$_2$/(±45)	181	211.040	16.6	195.74	8.1	195.09	7.8

As can be seen from Table 5.3, predictions treating the impact damage as a hole are within 17%. Treating impact damage as a delamination or as a region of reduced stiffness gives predictions within 14%, with the reduced stiffness model being slightly better. In general, these methods give reasonable predictions and, considering how easy they are to use, they are good for preliminary design.

5.7 Improved Model for Impact Damage Analysed as a Region of Reduced Stiffness

The discussion in Section 5.6 suggested that, while modelling impact damage as a region of reduced stiffness may give reasonably accurate predictions (see Table 5.3), it suffers from several limitations such as not knowing *a priori* the shape of the damage region and not knowing the stiffness distribution within the damage region. An improved model based on [14, 15] with even better accuracy for monolithic laminates is presented in this section.

5.7.1 Type and Extent of Damage for Given Impact Energy

In order to create an efficient and accurate model for CAI, one must first determine the extent of damage created during impact at a given energy level. It will be assumed throughout the rest of this chapter that the conditions of low-speed impact damage hold. This allows treatment of the problem as a quasi-static problem. Conceptually, the approach is as follows:

- For a given impact energy, determine the peak force applied and the area over which it acts.
- Determine the stresses everywhere in the laminate when the peak force acts.
- Use these stresses and some combination of in-plane and out-of-plane failure criteria to determine where through the thickness there is failure and what type it is: matrix cracks, delaminations, fibre breakage.

5.7.1.1 Peak Force during Impact

An energy balance is used. This follows the approaches proposed by Cairns [16] and Olsson [17]. It is assumed that the impact force reaches its peak value the moment the impacted plate reaches its maximum out-of-plane deflection and that both the plate and the impactor are, momentarily, stationary before they start springing back. It is also assumed that the total kinetic energy of the impactor at that moment has been transformed to strain energy in the plate and energy in locally indenting the plate. The energy balance then has the form

$$\frac{1}{2}m_i v_i^2 = U_p + E_{ind} \tag{5.6}$$

where the subscript 'i' refers to the impactor, U_p is the strain energy in the plate and E_{ind} is the indentation energy stored in indenting the plate. It should be noted that any energy lost in damaging the plate during impact is, at this point, neglected.

The energy stored in the plate can be equated to the work done for a linear system, which is what is assumed here as a first-order approximation:

$$U_p = W_p = \int_0^{w_{max}} F dw \tag{5.7}$$

where w is the out-of-plane deflection at the centre of the plate resulting from the impact force F. Note that F is not constant but varies with w.

For the relation between the force F and the deflection w, the contact force is assumed to be a point force. This is an approximation that will be reasonably accurate as long as the plate dimensions are significantly larger than the contact area. Then, the relationship between F and w is obtained by solving for the out-of-plane deflections w of a plate under a point force. The solution is, of course, a function of the boundary conditions. For the case of simply-supported plate edges, the centre deflection w_{max} of a rectangular plate with length a and width b is found to be [18]

$$w_{max} = \sum \sum \frac{(4F/ab)\sin^2(m\pi/2)\sin^2(n\pi/2)}{D_{11}(m\pi/a)^4 + 2(D_{12} + 2D_{66})(m^2 n^2 \pi^4/a^2 b^2) + D_{22}(n\pi/b)^4} \tag{5.8}$$

where D_{ij} refer to the bending stiffness matrix properties of the plate. Note that in this derivation it was assumed that $D_{16} = D_{26} = 0$. The double summation is over the number of terms needed for convergence (typically, $\max(m,n) = 25$). Equation 5.8 can be written in the form

$$F = kw_{max} \tag{5.9}$$

with

$$k = \frac{1}{\displaystyle\sum \sum \frac{(4/ab)\sin^2(m\pi/2)\sin^2(n\pi/2)}{(D_{11}(m\pi/a)^4 + 2(D_{12} + 2D_{66})(m^2 n^2 \pi^4/a^2 b^2) + D_{22}(n\pi/b)^4)}} \tag{5.10}$$

Substituting Equation 5.9 in Equation 5.7 permits evaluation of the strain energy of the plate in the form

$$U_p = \int_0^{w_{max}} kw dw = \frac{1}{2}kw_{max}^2 = \frac{F_{peak}^2}{2k} \tag{5.11}$$

For the indentation energy, an analogous relation between the force and the indentation δ is needed. This is obtained by assuming a Hertzian contact between the impactor and the plate. Hertz [19] showed that, as long as there is no damage in the plate or the impactor, the contact force is related to the indentation via

$$F = k_{\text{ind}}\delta^{\frac{3}{2}} \tag{5.12}$$

where the indentation constant k_{ind} is obtained for quasi-isotropic laminates, from [20, 21]

$$k_{\text{ind}} = \frac{4\sqrt{R}}{3\pi(K_1 + K_2)} \tag{5.13}$$

with R the impactor radius and K_1 and K_2 are compliance properties of the impactor and the plate, respectively, given by

$$K_1 = \frac{1 - v_i^2}{\pi E_i} \tag{5.14}$$

$$K_2 = \frac{\sqrt{A_{22}}\sqrt{(\sqrt{A_{11}A_{22}} + G_{zr})^2 - (A_{12} + G_{zr})^2}}{2\pi\sqrt{G_{zr}(A_{11}A_{22} - A_{12}^2)}} \tag{5.15}$$

with A_{11}, A_{22} and A_{12} stiffness properties for the impacted laminate given by

$$A_{11} = E_z(1 - v_{r\theta})\beta$$

$$A_{22} = \frac{E_r\beta(1 - v_{rz}^2\alpha)}{1 + v_{r\theta}}$$

$$A_{12} = E_r v_{rz}\beta$$

$$\beta = \frac{1}{1 - v_{r\theta} - 2v_{rz}^2\alpha}$$

$$\alpha = \frac{E_r}{E_z}$$

The indentation energy is then obtained as

$$E_{\text{ind}} = \int_0^{\delta_{\text{max}}} F d\delta = \frac{2}{5}k_{\text{ind}}\delta_{\text{max}}^{\frac{5}{2}} = \frac{2}{5}\frac{F^{\frac{5}{3}}}{k_{\text{ind}}^{\frac{2}{3}}} \tag{5.16}$$

where Equation 5.12 was used.

Equations 5.16 and 5.11 can be substituted in Equation 5.6 to obtain the total energy stored in the laminate in the form of strain energy and indentation energy. The peak force corresponding to a certain energy level is then obtained as follows:

- Select a value of F_{peak}.
- Use Equation 5.11 to obtain the strain energy U_p.

- Use Equation 5.16 to obtain the indentation energy E_{ind}.
- Substitute in Equation 5.6 to obtain the total energy.
- If the right-hand side of Equation 5.6 does not match the impact energy on the left-hand side, adjust F_{peak} accordingly and repeat the process until the left- and right-hand sides of Equation 5.6 are equal to each other within a preset tolerance.

Three points deserve further discussion:

1. The boundary conditions used for the plate under point load can make a significant difference. Repeating the procedure using clamped boundary conditions instead of simply-supported resulted in differences in w_{max} approaching 40%.
2. As already suggested, the Hertzian contact assumed in Equation 5.12 is not valid once damage starts. For more accurate results, Equation 5.12 should be replaced by a different model after onset of damage. For example, Talagani [22] has shown that the contact models by Christoforou and Yigit [23] and Yang and Sun [24] can be combined in a single model that matches experimental results very well even after damage onset.
3. The model presented here neglects the effect of the damage on the peak force. It would predict a higher force than would be measured during an impact test. This discrepancy is not as limiting as it appears. It will be discussed further in Section 5.7.1.3.

5.7.1.2 Stress Determination as a Result of the Peak Impact Force

For the stress determination, the contact force is no longer a point force. A point force is unrealistically conservative in this case. Instead, the Hertzian contact pressure distribution is used:

$$p = \frac{3F_{peak}}{2\pi R_c^2} \sqrt{1 - \frac{r^2}{R_c^2}} \tag{5.17}$$

Equation 5.17 describes the pressure exerted during impact of a spherical impactor with a plate when the impact force is F_{peak}. The contact radius is R_c, and r is the radial coordinate from the centre of impact. The contact pressure p is maximum at the centre of impact ($r = 0$) and 0 for $r > R_c$. The situation is shown in Figure 5.16.

With reference to Figure 5.16, the contact radius R_c is found to be

$$R_c = \sqrt{R^2 - (R - \delta_{max})^2} \tag{5.18}$$

where R is the impactor radius and δ_{max} is the maximum indentation obtained from Equation 5.12. Equation 5.18 is valid for a rigid impactor. Here, the impactor is assumed to be made out of steel whose stiffness is much higher than the stiffness of the composite plate; so this assumption is justified.

Given the applied pressure p, the stresses must be determined everywhere in the plate. The plate is assumed to be transversely isotropic; that is, it has the same stiffness

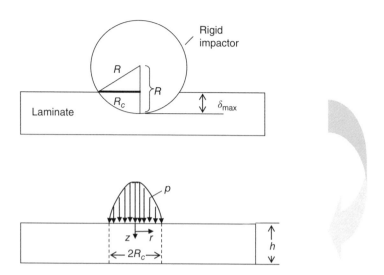

Figure 5.16 Impact geometry and contact pressure

in every in-plane direction, which, in general, will be different from the out-of-plane stiffness.

For convenience, cylindrical coordinates will be used. The assumptions on the applied load and stiffness of the plate lead to the conclusion that there will be no dependence on the θ direction. Furthermore, because of symmetry, the shear stresses $\tau_{r\theta}$ and $\tau_{\theta z}$ are zero. The stresses to be determined are therefore σ_r, σ_θ, σ_z and τ_{rz}, with z the out-of-plane direction.

The plate is divided in two regions. One is the region below the impact site, $0 \leq r \leq R_c$, and the other is the outboard region, $r > R_c$. The solution starts in the centre region and proceeds outwards. Inside the contact region, $0 \leq r \leq R_c$, the out-of-plane normal stress σ_z is assumed to have the form

$$\sigma_z = \frac{3F}{2\pi R_c^2}\sqrt{1 - \frac{r^2}{R_c^2}}\left[A_1 e^{-\phi_1\frac{z}{h}} + A_2 e^{\phi_1\frac{z}{h}} + A_3 e^{-\phi_2\frac{z}{h}} + A_4 e^{\phi_2\frac{z}{h}}\right] \tag{5.19}$$

By comparing with Equation 5.17 it can be seen that the r dependence is identical to the applied impact pressure p. The z dependence is in terms of four exponentials with two different unknown exponents. This is an assumption based on stress solutions obtained by applying energy minimization and variational calculus to a certain class or problems in composites [25, 26]. In such problems, the governing differential equation contains even derivatives of the unknown stress dependence and leads to exponentials in the form of those in Equation 5.19. So it is anticipated that a variational formulation treating the z dependence in σ_z as an unknown would lead to an analogous differential equation and functional form for the solution.

The stress equilibrium equations in cylindrical coordinates are now invoked:

$$\frac{\partial \sigma_r}{\partial r} + \frac{1}{r}\frac{\partial \tau_{r\theta}}{\partial \theta} + \frac{\partial \tau_{rz}}{\partial z} + \frac{\sigma_r - \sigma_\theta}{r} = 0$$

$$\frac{\partial \tau_{r\theta}}{\partial r} + \frac{1}{r}\frac{\partial \sigma_\theta}{\partial \theta} + \frac{\partial \tau_{\theta z}}{\partial z} + \frac{2\tau_{r\theta}}{r} = 0$$

$$\frac{\partial \tau_{rz}}{\partial r} + \frac{1}{r}\frac{\partial \tau_{\theta z}}{\partial \theta} + \frac{\partial \sigma_z}{\partial z} + \frac{\tau_{rz}}{r} = 0 \qquad (5.20\text{a}-\text{c})$$

The assumptions of no dependence on θ and the fact that $\tau_{r\theta}$ and $\tau_{\theta z}$ are zero reduce the equilibrium equations to the following two:

$$\frac{\partial \sigma_r}{\partial r} + \frac{\partial \tau_{rz}}{\partial z} + \frac{\sigma_r - \sigma_\theta}{r} = 0$$

$$\frac{\partial \tau_{rz}}{\partial r} + \frac{\partial \sigma_z}{\partial z} + \frac{\tau_{rz}}{r} = 0 \qquad (5.21\text{a}-\text{b})$$

The expression for σ_z, Equation 5.19, can be substituted into Equation 5.21b, which can then be solved for the shear stress τ_{rz}. This is done by multiplying Equation 5.21b by r, which would make the terms involving τ_{rz} become a partial derivative:

$$r\frac{\partial \tau_{rz}}{\partial r} + \tau_{rz} = \frac{\partial}{\partial r}(r\tau_{rz})$$

allowing easy integration to obtain

$$\tau_{rz} = \frac{F}{2\pi r}\left[\left(1 - \frac{r^2}{R_c^2}\right)^{\frac{3}{2}} - 1\right]$$

$$\cdot \left[-A_1\frac{\phi_1}{h}e^{-\varphi_1\frac{z}{h}} + A_2\frac{\phi_1}{h}e^{\varphi_1\frac{z}{h}} - A_3\frac{\phi_2}{h}e^{-\varphi_2\frac{z}{h}} + A_4\frac{\phi_2}{h}e^{\varphi_2\frac{z}{h}}\right] \qquad (5.22)$$

To arrive at Equation 5.22, the condition that $\tau_{rz} = 0$ at $r = 0$ was also imposed. The following boundary conditions are now used in the contact region:

$$\sigma_z(z = 0) = -p$$

$$\sigma_z(z = h) = 0$$

$$\tau_{rz}(z = 0) = 0$$

$$\tau_{rz}(z = h) = 0 \qquad (5.23\text{a}-\text{d})$$

These conditions state that, at the top of the laminate ($z = 0$), the normal stress equals the applied pressure p given by Equation 5.17. The minus sign on the right-hand side of Equation 5.23a indicates that the applied stress is compressive. At the same location, the shear stress is zero. At the bottom of the laminate ($z = h$), both the normal stress σ_z and shear stress τ_{rz} are zero. Using Equations 5.19 and 5 to substitute in Equation 5.23a–d gives four equations in the four unknowns A_1–A_4:

$$\begin{bmatrix} 1 & 1 & 1 & 1 \\ e^{-\varphi_1} & e^{\varphi_1} & e^{-\varphi_2} & e^{\varphi_2} \\ -\varphi_1 & \varphi_1 & -\varphi_2 & \varphi_2 \\ -\varphi_1 e^{-\varphi_1} & \varphi_1 e^{\varphi_1} & -\varphi_2 e^{-\varphi_2} & \varphi_2 e^{\varphi_2} \end{bmatrix} \begin{Bmatrix} A_1 \\ A_2 \\ A_3 \\ A_4 \end{Bmatrix} = \begin{Bmatrix} -1 \\ 0 \\ 0 \\ 0 \end{Bmatrix} \qquad (5.24a–d)$$

Equations 5.24a–d permit determination of A_1–A_4 if the exponents φ_1 and φ_2 are known. The latter are determined by minimising the energy of the laminate. But in order to do this, the remaining stresses in the contact region and all the stresses in the outboard region $r > R_c$ must be determined.

In order to determine the in-plane stresses σ_r and σ_θ, the stress–strain equations are used:

$$\begin{Bmatrix} \sigma_r \\ \sigma_\theta \\ \sigma_z \\ \tau_{rz} \end{Bmatrix} = \begin{bmatrix} C_{rr} & C_{r\theta} & C_{rz} & 0 \\ C_{r\theta} & C_{\theta\theta} & C_{\theta z} & 0 \\ C_{rz} & C_{\theta z} & C_{zz} & 0 \\ 0 & 0 & 0 & C_{55} \end{bmatrix} \begin{Bmatrix} \varepsilon_r \\ \varepsilon_\theta \\ \varepsilon_z \\ \gamma_{rz} \end{Bmatrix} \qquad (5.25a–d)$$

For a transversely isotropic material, $C_{rr} = C_{\theta\theta}$ and $C_{rz} = C_{\theta z}$. For the same reason, which means there is no dependence on the θ coordinate, the strain displacement equations

$$\varepsilon_r = \frac{\partial u_r}{\partial r}$$

$$\varepsilon_\theta = \frac{1}{r}\frac{\partial u_\theta}{\partial \theta} + \frac{u_r}{r}$$

simplify to

$$\varepsilon_r = \frac{\partial u_r}{\partial r}$$

$$\varepsilon_\theta = \frac{u_r}{r} \qquad (5.26a–b)$$

Equation 5.26b is multiplied by r and then differentiated with respect to r:

$$\frac{\partial(r\varepsilon_{\theta\theta})}{\partial r} = \frac{\partial u_r}{\partial r} \qquad (5.27)$$

$$\varepsilon_r = \frac{\partial(r\varepsilon_{\theta\theta})}{\partial r} \qquad (5.28)$$

This can be substituted in Equation 5.26a to eliminate u_r.

Equations 5.25a–d, 5.21a and 5.28 can now be combined to yield a relationship between σ_r and τ_{rz}:

$$\frac{\partial}{\partial r}\left(r^3\frac{\partial\sigma_r}{\partial r}\right) + C_1 r^2\frac{\partial\tau_{rz}}{\partial z} + C_2 r^2\frac{\partial\sigma_{zz}}{\partial r} + r^3\frac{\partial^2\tau_{rz}}{\partial r\partial z} = 0 \qquad (5.29)$$

with

$$C_1 = \frac{2(C_{rr}C_{zz} + (1/2)C_{r\theta}C_{zz} - (3/2)C_{rz}^2)}{C_{rr}C_{zz} - C_{rz}^2}$$

$$C_2 = -\frac{C_{rz}(C_{rr} - C_{r\theta})}{C_{rr}C_{zz} - C_{rz}^2}$$

Equation 5.29 can be solved for σ_r because τ_{rz} is known from Equation 5.22. Then, with σ_r known, substitution in Equation 5.21a will give σ_θ. The result is

$$
\sigma_r = -\frac{F}{2\pi}\frac{C_{rz}(C_{rr}-C_{r\theta})}{(C_{rr}C_{zz}-C_{rz}^2)}\left[\frac{1}{r^2}\left(\sqrt{1-\frac{r^2}{R_c^2}}-1\right)-\frac{1}{R_c^2}\sqrt{1-\frac{r^2}{R_c^2}}\right]
$$

$$
\cdot\left[A_1 e^{-\varphi_1\frac{z}{h}}+A_2 e^{\varphi_1\frac{z}{h}}+A_3 e^{-\varphi_2\frac{z}{h}}+A_4 e^{\varphi_2\frac{z}{h}}\right]
$$

$$
+\frac{3F}{2\pi R_c^2}\left[-\frac{3R_c^4}{90r^2}\left\{2\left(1-\sqrt{1-\frac{r^2}{R_c^2}}\right)+\left(\sqrt{1-\frac{r^2}{R_c^2}}-1\right)\right\}\right.
$$

$$
-\frac{R_c^2}{6\Lambda_1}\ln\left(1+\sqrt{1-\frac{r^2}{R_c^2}}\right)+(14\Lambda_1+12)\frac{R_c^2}{90}\sqrt{1-\frac{r^2}{R_c^2}}-\frac{2r^2}{90}\left(1+\frac{1}{\Lambda_1}\right)
$$

$$
\left.\sqrt{1-\frac{r^2}{R_c^2}}\right]\cdot\left[\frac{\varphi_1^2}{h^2}A_1 e^{-\varphi_1\frac{z}{h}}+\frac{\varphi_1^2}{h^2}A_2 e^{\varphi_1\frac{z}{h}}+\frac{\varphi_2^2}{h^2}A_3 e^{-\varphi_2\frac{z}{h}}+\frac{\varphi_2^2}{h^2}A_4 e^{\varphi_2\frac{z}{h}}\right]\quad(5.30)
$$

$$
\sigma_\theta = -\frac{F}{2\pi}\frac{C_{rz}(C_{rr}-C_{r\theta})}{(C_{rr}C_{zz}-C_{rz}^2)}\left[\frac{1}{r^2}\left(1-\sqrt{1-\frac{r^2}{R_c^2}}\right)+\frac{1}{\sqrt{1-\frac{r^2}{R_c^2}}}\left(\frac{2r^2}{R_c^4}-\frac{1}{R_c^2}\right)\right]
$$

$$
\cdot\left[A_1 e^{-\varphi_1\frac{z}{h}}+A_2 e^{\varphi_1\frac{z}{h}}+A_3 e^{-\varphi_2\frac{z}{h}}+A_4 e^{\varphi_2\frac{z}{h}}\right]
$$

$$
+\frac{F}{2\pi}\left\{\frac{R_c^2}{30r^2}\left(6-\frac{3}{\Lambda_1}+\left(6-\frac{9}{\Lambda_1}\right)\sqrt{1-\frac{r^2}{R_c^2}}\right)+\frac{1}{2\Lambda_1}\ln\left(1+\sqrt{1-\frac{r^2}{R_c^2}}\right)\right.
$$

$$
-\frac{1}{30\Lambda_1}\sqrt{1-\frac{r^2}{R_c^2}}\left(14+12\Lambda_1-24\Lambda_1\frac{r^2}{R_c^2}\right)-\frac{1}{30\Lambda_1\sqrt{1-\frac{r^2}{R_c^2}}}
$$

$$
\cdot\left(3+6\Lambda_1+\frac{15r^2}{R_c^2\left(1+\sqrt{1-\frac{r^2}{R_c^2}}\right)}-14\frac{r^2}{R_c^2}-12\Lambda_1\frac{r^2}{R_c^2}+2\frac{r^4}{R_c^4}+6\Lambda_1\frac{r^4}{R_c^4}\right)
$$

$$
\left.+\left(\sqrt{1-\frac{r^2}{R_c^2}}-1\right)^3\right\}\cdot\left[\frac{\varphi_1^2}{h^2}A_1 e^{-\varphi_1\frac{z}{h}}+\frac{\varphi_1^2}{h^2}A_2 e^{\varphi_1\frac{z}{h}}\right.
$$

$$
\left.+\frac{\varphi_2^2}{h^2}A_3 e^{-\varphi_2\frac{z}{h}}+\frac{\varphi_2^2}{h^2}A_4 e^{\varphi_2\frac{z}{h}}\right]\quad(5.31)
$$

with Λ_1 given by

$$\Lambda_1 = \frac{(C_{rr} - C_{r\theta})(C_{rr}C_{zz} - C_{rz}^2)}{C_{rr}^2 C_{zz} - 2C_{rr}C_{rz}^2 - C_{r\theta}^2 C_{zz} + 2C_{r\theta}C_{rz}^2}$$

Equations 5.30 and 5.31 complete the stress determination in the contact region. The solution now proceeds with the determination of stresses in the outboard region ($r > R_c$). The approach is similar. An expression is assumed for the σ_z stress, and the remaining stresses are obtained by using stress equilibrium and satisfying the boundary conditions. These conditions now include matching of the stresses between the two regions (contact and outboard) at their interface ($r = R_c$).

The expression for σ_z must be zero at the top and bottom surface of the laminate. It must also be zero at the far field ($r \to \infty$) because the effects of the impact, which give rise to the interlaminar stresses σ_z and τ_{rz}, die out far from the impact site. For this reason, the following expression is assumed for σ_z:

$$\sigma_{zo} = -\frac{3F}{2\pi R_c^2} e^{\psi(R_c - r)}$$

$$\cdot \left[A_1 e^{-\phi_1 \frac{z}{h}} + A_2 e^{\phi_1 \frac{z}{h}} + A_3 e^{-\phi_2 \frac{z}{h}} + A_4 e^{\phi_2 \frac{z}{h}} + 1 - 3\left(\frac{z}{h}\right)^2 + 2\left(\frac{z}{h}\right)^3 \right] \quad (5.32)$$

where the subscript 'o' denotes the outboard region.

The exponential in r guarantees that σ_z decays to zero at large r (provided the unknown exponent ψ is positive). Also, in addition to the four exponentials in z which are needed to match σ_r and τ_{rz} at $r = R_c$, a cubic polynomial in z is added which is needed to reproduce the standard quadratic distribution through the thickness for the shear stress τ_{rz} for large values of r.

Equation 5.32 can be substituted in Equation 5.21b, and the resulting equation is solved for the shear stress τ_{rz}. Requiring that $\tau_{rz}(r = R_c) = \tau_{rzo}(r = R_c)$ and that τ_{rzo} be zero at the top and bottom surfaces of the laminate gives

$$\tau_{rzo} = -\frac{3F}{2\pi R_c^2} e^{\psi(R_c - r)} \frac{(\psi r + 1)}{\psi^2 r}$$

$$\cdot \left[-A_1 \frac{\phi_1}{h} e^{-\phi_1 \frac{z}{h}} + A_2 \frac{\phi_1}{h} e^{\phi_1 \frac{z}{h}} - A_3 \frac{\phi_2}{h} e^{-\phi_2 \frac{z}{h}} + A_4 \frac{\phi_2}{h} e^{\phi_2 \frac{z}{h}} - 6\frac{z}{h^2} + 6\frac{z^2}{h^3} \right]$$

$$+ \frac{3F}{\pi r} \left(\frac{z^2}{h^3} - \frac{z}{h^2} \right) \quad (5.33)$$

with

$$\psi = \frac{3}{2R_c} + \frac{\sqrt{21}}{2R_c}$$

Now, using Equations 5.29 and 5.21a, along with the condition that σ_{ro} and σ_r be the same at $r = R_c$, gives

$$\sigma_{rro} = \frac{F}{2\pi} \frac{C_{rz}(C_{rr} - C_{r\theta})}{(C_{rr}C_{zz} - C_{rz}^2)} \left(\frac{1}{r^2} - \frac{(1 - (R_c^2/r^2))}{R_p^2(1 - (R_c^2/R_p^2))} \right)$$

$$\cdot \left(A_1 e^{-\varphi_1 \frac{z}{h}} + A_2 e^{\varphi_1 \frac{z}{h}} + A_3 e^{-\varphi_2 \frac{z}{h}} + A_4 e^{\varphi_2 \frac{z}{h}} \right)$$

$$- \frac{FR_c^2}{120\pi} \left(12 - \frac{6}{\Lambda_1} \right) \left(\frac{1}{r^2} - \frac{(1 - (R_c^2/r^2))}{R_p^2(1 - (R_c^2/R_p^2))} \right)$$

$$\cdot \left\{ \frac{\varphi_1^2}{h^2} A_1 e^{-\varphi_1 \frac{z}{h}} + \frac{\varphi_1^2}{h^2} A_2 e^{\varphi_1 \frac{z}{h}} + \frac{\varphi_2^2}{h^2} A_3 e^{-\varphi_2 \frac{z}{h}} + \frac{\varphi_2^2}{h^2} A_4 e^{\varphi_2 \frac{z}{h}} \right\}$$

$$+ \frac{3F}{2\pi R_c^2 \psi^2} \frac{C_{rz}(C_{rr} - C_{r\theta})}{(C_{rr}C_{zz} - C_{rz}^2)} \left\{ \frac{1}{r^2} (\psi R_c + 1 - (\psi r + 1) e^{\psi(R_c - r)}) \right.$$

$$- \frac{(1 - (R_c^2/r^2))}{R_p^2(1 - (R_c^2/R_p^2))} (\psi R_c + 1 - (\psi R_p + 1) e^{\psi(R_c - R_p)})$$

$$\cdot \left\{ A_1 e^{-\varphi_1 \frac{z}{h}} + A_2 e^{\varphi_1 \frac{z}{h}} + A_3 e^{-\varphi_2 \frac{z}{h}} + A_4 e^{\varphi_2 \frac{z}{h}} + 1 - 3 \left(\frac{z}{h} \right)^2 + 2 \left(\frac{z}{h} \right)^3 \right\}$$

$$+ \frac{3F}{2\pi \Lambda_1 R_c^2 \psi^4} \left\{ \frac{1}{r^2} \left(e^{\psi(R_c - r)} \left(\frac{3}{2} - 3\Lambda_1 - 3r \left(\Lambda_1 - \frac{1}{2} \right) - \Lambda_1^2 \psi^2 r^2 \right) \right. \right.$$

$$+ \frac{1}{2} R_c^2 \psi^2 e^{\psi R_c} Ei(\psi R_c) - \frac{1}{2} \Lambda_1 R_c^2 e^{\psi R_c} Ei(\psi r) \psi^4 r^2 + 3 \left(\Lambda_1 - \frac{1}{2} \right)$$

$$\cdot \psi R_c - \frac{3}{2} + 3\Lambda_1 \right) - \frac{(1 - (R_c^2/r^2))}{R_p^2(1 - (R_c^2/R_p^2))} \left(e^{\psi(R_c - R_p)} \left(\frac{3}{2} - 3\Lambda_1 - 3R_p \left(\Lambda_1 - \frac{1}{2} \right) \right. \right.$$

$$\left. - \Lambda_1^2 \psi^2 R_p^2 \right) + \frac{1}{2} R_c^2 \psi^2 e^{\psi R_c} Ei(\psi R_c) - \frac{1}{2} \Lambda_1 R_c^2 e^{\psi R_c} Ei(\psi R_p) \psi^4 R_p^2$$

$$\left. \left. + 3 \left(\Lambda_1 - \frac{1}{2} \right) \psi R_c - \frac{3}{2} + 3\Lambda_1 \right) \right\}$$

$$\cdot \left\{ \frac{\varphi_1^2}{h^2} A_1 e^{-\varphi_1 \frac{z}{h}} + \frac{\varphi_1^2}{h^2} A_2 e^{\varphi_1 \frac{z}{h}} + \frac{\varphi_2^2}{h^2} A_3 e^{-\varphi_2 \frac{z}{h}} + \frac{\varphi_2^2}{h^2} A_4 e^{\varphi_2 \frac{z}{h}} - \frac{6}{h^2} + 12 \frac{z}{h^3} \right\}$$

$$+ \frac{1}{2\Lambda_1} \left[\frac{\ln(R_c) R_c^2}{r^2} - \ln(r) - \frac{(1 - (R_c^2/r^2))}{(1 - (R_c^2/R_p^2))} \left(\frac{\ln(R_c) R_c^2}{R_p^2} - \ln(R_p) \right) \right]$$

$$\cdot \left\{ \frac{3F}{\pi} \left(\frac{2z}{h^3} - \frac{1}{h^2} \right) \right\} \tag{5.34}$$

$$\sigma_{\theta\theta o} = \frac{F}{2\pi} \frac{C_{rz}(C_{rr} - C_{r\theta})}{(C_{rr}C_{zz} - C_{rz}^2)} \left(\frac{3}{r^2} - \frac{(1 - (R_c^2/r^2))}{R_p^2(1 - (R_c^2/R_p^2))} - \frac{r^2 + R_c^2}{r^2(R_c^2 - R_p^2)} \right)$$

$$\cdot \left(A_1 e^{-\varphi_1 \frac{z}{h}} + A_2 e^{\varphi_1 \frac{z}{h}} + A_3 e^{-\varphi_2 \frac{z}{h}} + A_4 e^{\varphi_2 \frac{z}{h}} \right)$$

$$+ \frac{FR_c^2}{120\pi} \left(12 - \frac{6}{\Lambda_1} \right) \left(\frac{1}{r^2} + \frac{(1 - (R_c^2/r^2))}{R_p^2(1 - (R_c^2/R_p^2))} + \frac{2(r^2 + R_c^2)}{r^2(R_c^2 - R_p^2)} \right)$$

$$\cdot \left\{ \frac{\varphi_1^2}{h^2} A_1 e^{-\varphi_1 \frac{z}{h}} + \frac{\varphi_1^2}{h^2} A_2 e^{\varphi_1 \frac{z}{h}} + \frac{\varphi_2^2}{h^2} A_3 e^{-\varphi_2 \frac{z}{h}} + \frac{\varphi_2^2}{h^2} A_4 e^{\varphi_2 \frac{z}{h}} \right\}$$

$$- \frac{3F}{2\pi R_c^2} \frac{C_{rz}(C_{rr} - C_{r\theta})}{(C_{rr}C_{zz} - C_{rz}^2)} \frac{1}{r^2} \left(\frac{\psi R_c + 1}{\psi^2} - \frac{\psi r + 1}{\psi^2} e^{\psi(R_c - r)} - r \frac{e^{\psi(R_c - r)}}{\psi} \right.$$

$$+ r \frac{(\psi r + 1)e^{\psi(R_c - r)}}{\psi} + \frac{2(\psi r + 1)e^{\psi(R_c - r)}}{\psi^2} - \frac{2(\psi R_c + 1)e^{\psi(R_c - r)}}{\psi^2}$$

$$+ \frac{1}{\psi^2} \frac{(r^2 + R_c^2)}{(R_c^2 - R_p^2)} \left(\psi R_c + 1 - (\psi R_p + 1) e^{\psi(R_c - r)} \right) \right)$$

$$\cdot \left\{ A_1 e^{-\varphi_1 \frac{z}{h}} + A_2 e^{\varphi_1 \frac{z}{h}} + A_3 e^{-\varphi_2 \frac{z}{h}} + A_4 e^{\varphi_2 \frac{z}{h}} + 1 - 3 \left(\frac{z}{h} \right)^2 + 2 \left(\frac{z}{h} \right)^3 \right\}$$

$$+ \frac{3F}{2\pi\Lambda_1 \psi^4 R_c^2 r^2} \left[e^{\psi(R_c - r)} \left(\frac{3}{2} - 3\Lambda_1 - 3 \left(\Lambda_1 - \frac{1}{2} \right) \psi r - \Lambda_1 \psi^3 r^3 - 2\Lambda_1 \psi^2 r^2 \right) \right.$$

$$+ \frac{1}{2} R_c^2 \psi^2 e^{\psi R_c} Ei(\psi R_c) - \frac{1}{2} R_c^2 e^{\psi R_c} Ei(\psi r)\psi^2 r^2 + \Lambda_1 \psi^2 R_c^2$$

$$+ 3 \left(\Lambda_1 - \frac{1}{2} \right) \psi R_c + 3\Lambda_1 - \frac{3}{2} + \frac{(r^2 + R_c^2)}{(R_c^2 - R_p^2)}$$

$$\cdot \left(e^{\psi(R_c - R_p)} \left(\frac{3}{2} - 3\Lambda_1 - 3 \left(\Lambda_1 - \frac{1}{2} \right) \psi R_p - \Lambda_1 \psi^2 R_p^2 \right) + \frac{1}{2} R_c^2 \psi^2 e^{\psi R_c} Ei(\psi R_c) \right.$$

$$- \frac{\Lambda_1}{2} e^{\psi R_c} Ei(\psi R_p)\psi^2 R_c^2 R_p^2 + \Lambda_1 \psi^2 R_c^2 + 3 \left(\Lambda_1 - \frac{1}{2} \right) \psi R_c + 3\Lambda_1 - \frac{3}{2} \right) \right]$$

$$\cdot \left\{ \frac{\varphi_1^2}{h^2} A_1 e^{-\varphi_1 \frac{z}{h}} + \frac{\varphi_1^2}{h^2} A_2 e^{\varphi_1 \frac{z}{h}} + \frac{\varphi_2^2}{h^2} A_3 e^{-\varphi_2 \frac{z}{h}} + \frac{\varphi_2^2}{h^2} A_4 e^{\varphi_2 \frac{z}{h}} - \frac{6}{h^2} + 12 \frac{z}{h^3} \right\} +$$

$$+ \left(\frac{3F}{2\pi\Lambda_1 r^2} \left(2r^2 \left\{ \frac{\ln(r)}{2} + \Lambda_1 - \frac{1}{2} \right\} - R_c^2 \ln(R_c) + \frac{R_p^2(r^2 + R_c^2)}{(R_c^2 - R_p^2)} \right. \right.$$

$$\left. \left. \cdot \left[\frac{\ln(R_c) R_c^2}{R_p^2} - \ln(R_p) \right] \right) \right) \cdot \left[\frac{2z}{h^3} - \frac{1}{h^2} \right]$$

(5.35)

with R_p the radius to the outer edge of the specimen and E_i the exponential integral

$$E_i(x) = \int_{-x}^{\infty} \frac{e^{-t}}{t} dt$$

At this point, only the exponents φ_1 and φ_2 are undetermined. They are found by minimising the energy of the entire plate:

$$\Pi = \frac{1}{2} \int\!\!\int\!\!\int \underline{\sigma}^T \underline{S} \underline{\sigma} dV - \int\!\!\int \underline{T}^T \underline{u} dA \qquad (5.36)$$

where $\underline{u}$ denotes prescribed displacements and $\underline{T}$ the corresponding tractions. An underscore in Equation 5.36 denotes a matrix. $\underline{S}$ is the compliance tensor which can be obtained by inverting the stiffness matrix in Equations 5.25a–d.

As there are no prescribed displacements, the second term in Equation 5.36 vanishes. Then, evaluating the integrals for the first term and setting

$$\frac{\partial \Pi}{\partial \varphi_1} = 0 \qquad (5.37)$$

$$\frac{\partial \Pi}{\partial \varphi_2} = 0 \qquad (5.38)$$

gives two polynomial equations in the two unknowns φ_1 and φ_2. These are 10th order polynomials and can only be solved by iteration. The scheme used here is based on minimising the residual error for the two equations using a gradient approach. Typically, very few iterations are needed to solve Equations 5.37 and 5.38. Of the multiple solutions, the one minimising the energy is selected. In general, some of the φ_1 and φ_2 pairs may be complex. For all the cases run so far, the φ_1, φ_2 values that minimised the energy were both real.

One interesting characteristic of the solution just derived is how the shear stress τ_{rz} varies as a function of z at different r locations away from the impact site. Typical results are shown in Figure 5.17.

It can be seen from Figure 5.17 that, under the impact site and close to it, the distribution of τ_{rz} is not symmetric but reaches its peak value closer to the impacted surface. Further away from the impact site, in the outboard region, it tends to recover the well-known symmetric (quadratic) distribution through the thickness. This will have some implications related to onset of damage predictions, which will be discussed in Section 5.7.1.3.

Accuracy of the Determined Stresses

Before proceeding with failure predictions, it is important to validate the solution just presented. The stresses will be compared with other solutions in order to gain confidence in their accuracy.

The first comparison will be done for isotropic thick plates. Using stress functions, Love [27] developed the solution for the stresses in a plate of infinite thickness under the pressure load p in Equation 5.17. The solution presented here is for plates of finite thickness, so it is expected to get closer to the solution by Love as h,

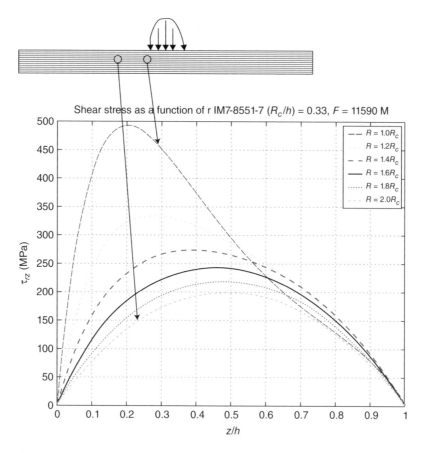

Figure 5.17 Through-the-thickness variation of τ_{rz} at different r locations

the plate thickness, increases. The shear stress τ_{rz} predicted by the present method for an isotropic plate with thickness $h = 10.0\,\text{mm}$, contact radius $R_c/h = 0.20$ and applied peak force $F_{\text{peak}} = 26,090\,\text{N}$ is compared with the solution by Love [27] in Figure 5.18. This is done at $r = R_c$, where the shear stress is maximum. The plate material has Young's modulus $E = 69\,\text{GPa}$ and Poisson's ratio $v = 0.3$.

As can be seen from Figure 5.18, the present solution is quite close to that obtained by Love. There are discrepancies, which are due to the fact that the plate has finite thickness and the solution by Love is for infinite plates. This is evident at the bottom surface of the plate, $z/h = 1$, where the present solution goes to zero, as it should, while the solution by Love is still finite. Love's solution has a higher maximum than the present solution. In order to maintain overall force equilibrium and integrate to the same total force, it has to cross the present solution and get below it over a range of z/h values. The area between the two curves over the region where Love's solution exceeds the present solution ($z/h < 0.15$) is approximately equal to that where Love's solution is below the present solution ($z/h > 0.15$); so both solutions integrate to the same shear force.

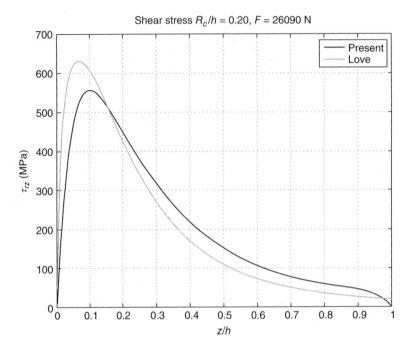

Figure 5.18   Predicted shear stress through the thickness compared to solution by Love (*See insert for colour representation of this figure.*)

The comparison of normal stress σ_z through the thickness for the same plate but at $r = 0$ where the normal stress is maximised is shown in Figure 5.19.

Again the two solutions are close. Love's solution does not go to zero at $z/h = 1$ because, as already mentioned, it is for a plate of infinite thickness. The present solution recovers exactly the boundary condition of zero stress at z/h. In this respect, the present solution is expected to be more accurate than Love's.

A more interesting comparison can be made against finite element (FE) results obtained for a quasi-isotropic plate of finite thickness. The FE results were obtained from [22] for a $[0/45/90/-45]_{2s}$ laminate with thickness 2.96 mm under a pressure load corresponding to a force 1030 N and $R_c/h = 0.28$.

The shear stress predicted by the present method is compared with the FE results from [22] in Figure 5.20.

There is good agreement between the two methods. There are discrepancies in the range $z/h < 0.5$, which are considered minor and probably due to the fact that the FE model is displacement-based and it has some difficulty satisfying the stress-free boundary condition point-wise at the top (and bottom) of the laminate.

The corresponding comparison for the normal stress σ_z is shown in Figure 5.21. The agreement in this case is even better than in Figure 5.20, and the discrepancies are small and confined to the region $0.15 < z/h < 0.35$.

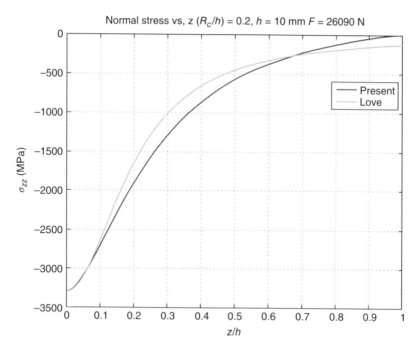

Figure 5.19 Predicted normal stress through the thickness compared to the solution by Love (*See insert for colour representation of this figure.*)

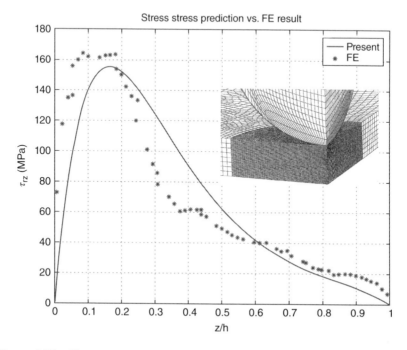

Figure 5.20 Shear stress at $r = R_c$ compared to FE for $[0/45/90/-45]_{2s}$ laminate

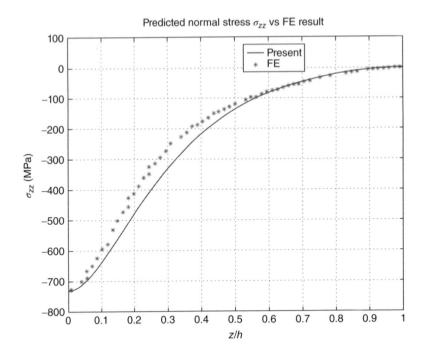

Figure 5.21   Normal stress at $r = 0$ compared to FE results from [22]

The results in Figures 5.18–5.21 suggest that the stresses below the impact site are captured accurately by the present method. However, there is a limiting case as the contact area becomes small and the pressure load approaches a point load, where the accuracy of the present method degrades. If $R_c/h < 0.15$, the stresses predicted by the present method depart from those predicted by FE. This is a direct result of the assumptions made for the z dependence of the σ_z stress. The cubic polynomial included in Equation 5.32 on one hand allows satisfying the boundary conditions at the top and bottom of the laminate and recovery of the quadratic distribution of the shear stress far from the impact site but, on the other, imposes a shape that is too rigid for small R_c/h values. However, as long as $R_c/h > 0.15$, the present solution is sufficiently accurate and can be used in a failure criterion to predict damage in the laminate in the vicinity of the impact site. This is done in the next section.

5.7.1.3   Determination of Type and Extent of Damage under the Impact Site

An idealised sketch of the damage under and near the impact site in a laminate is shown in Figure 5.22. This damage, in the form of matrix cracks, delaminations, broken fibres and local indentation, starts early during the impact event way before the peak force calculated in Section 5.7.1.1 is reached. It is important to note that, even though the laminate is quasi-isotropic in its plane and the loading during impact is

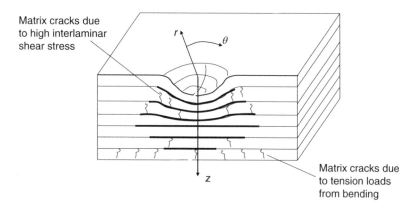

Matrix cracks due
to high interlaminar
shear stress

Matrix cracks due
to tension loads
from bending

Figure 5.22   Damage through the thickness of an impacted laminate

symmetric, the damage pattern is not necessarily symmetric. The transverse shear strength of the material is directional, meaning that failure due to shear stress exceeding the corresponding strength value will not be symmetric. This asymmetry will give rise to further asymmetry in failure as load is redistributed locally from failed plies to adjacent plies leading to localised failure in those plies. In addition, the delaminations created due to this asymmetry will reduce locally the bending stiffness of the laminate while under impact, leading to high localised bending strains, which promote further damage creation. Focusing only on delamination creation, while delaminations start mostly in mode II, the localised asymmetric bending introduces mode I fracture with different mode mixities as one goes around the impact site at a given distance r from the centre of impact.

It helps to visualise damage creation and growth better if one looks qualitatively at the distribution of the different stresses obtained in the previous section. The normal stress distribution σ_z is shown in Figure 5.23 at different r locations. It can be seen that this stress is always compressive. It is maximum at the top of the laminate, the region of contact between the impactor and the laminate, and reduces monotonically to zero at the bottom surface of the laminate. Furthermore, the magnitude of the stress decreases with increasing radial distance r from the centre of the impact site. This means that, near the impact site, high transverse compressive stresses are present and will lead to matrix and fibre failure through local compression and shear.

As a first approximation, one can predict failure in transverse compression by finding when the local stress exceeds the corresponding ultimate strength:

$$\frac{|\sigma_z|}{Y^c} = 1 \tag{5.39}$$

where, as an approximation to the out-of-plane compression strength, Y^c the in-plane transverse compression strength (perpendicular to the fibres) is used.

The corresponding notional plot for shear stress τ_{rz} is shown in Figure 5.24. The shear stress is zero at the top and bottom surfaces of the laminate and reaches a

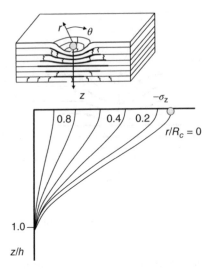

Figure 5.23   Normal stress distribution throughout impacted laminate

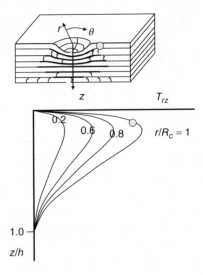

Figure 5.24   Transverse shear stress distribution throughout impacted laminate

maximum in between. Closer to the impact site, the location of the maximum moves towards the top surface, approximately at the one-quarter point. Further away from the impact site, the maximum of the shear stress occurs at the laminate mid-plane. This suggests that delaminations will first appear below the impact site near the top surface of the laminate. As the load increases, these delaminations become less effective because of excessive damage near the impacted surface. Other delaminations will appear near the laminate mid-plane as delaminations created at the edges of the contact area grow both outboard and towards the centre of the impact site. τ_{rz} is zero at the centre of the impact site ($r = 0$) and maximum at the edge of the contact region, $r = R_c$. With these in mind, a simple check for transverse shear failure would have the form

$$\frac{|\tau_{rz}|}{S_z} = 1 \qquad\qquad (5.40)$$

where S_z refers to the out-of-plane transverse shear strength. It is important to note that the value of S_z depends on the fibre orientation. Denoting by 1 the fibre direction and by 3 the out-of-plane direction, then, for typical carbon/epoxy unidirectional materials

$$S_{13} > S_{23}$$

This means that, depending on the way the local transverse shear stress is acting, the strength can be different and, for some materials, this difference can be as much as 30%. For this reason, the τ_{rz} stress at any r, θ, z location, is resolved in two shear stresses in the local ply coordinates to obtain its 13 and 23 components. These are then directly compared to the S_{13} and S_{23} values, respectively.

Typical distributions for σ_r and σ_θ are shown in Figure 5.25. It is interesting to note that σ_r may reach its highest value outside the contact region, for $r > R_c$. Also, σ_θ is discontinuous at $r = R_c$ and decays to zero for large r (or a constant value depending on the boundary conditions there). Both stresses may cause in-plane failure of the fibre of the matrix. Here, they were used in a maximum stress criterion.

The discussion so far focussed on predicting the onset of damage. It does not directly allow determination of damage growth. In principle, once damage starts, and depending on the type of damage, one would have to adjust stiffness and strength of damaged

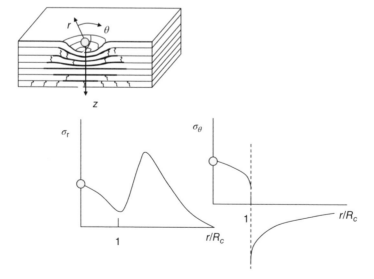

Figure 5.25   In-plane stresses in an impacted laminate

plies accordingly and then proceed with a progressive failure approach. While this would be more accurate, it would also be very intensive computationally and would most likely involve the use of finite elements. The purpose of the analysis developed here, however, is to come up with an efficient and reasonably accurate approach that can be used for quick evaluation of different design candidates and might help in formal optimization of laminates. For this reason, the damage size and type will be estimated using an approximation.

The approximation has its roots in the Whitney–Nuismer approach discussed in detail in Section 2.4. In that approach, the size of damage near the edge of a hole at failure is determined as a characteristic distance over which, when the stress is averaged, equals the undamaged strength of the material. This is based on the idea that, as soon as damage starts at the hole edge, the stress there will be limited to the un-notched strength of the laminate and will not increase further towards the value that the stress concentration at the hole edge would require. In this way, the elasticity solution can be used and, averaged over a characteristic distance, will give a good prediction of the laminate strength in the presence of a hole.

An analogous approach is used here. It is recognised that damage has started before the peak force during impact is reached. This damage will limit the stresses over the region where it occurred to their undamaged strength values. As a first approximation, then, by averaging the stresses over a distance such that the area under the averaged stress region equals the area under the elasticity solution in Section 5.7.1.2, the region over which a specific type of damage has occurred can be determined. This averaging process essentially guarantees that force equilibrium is maintained in an average sense. The situation is shown for the out-of-plane stresses in Figure 5.26. Maximum stress was used here, but other criteria can be used if they are considered more reliable.

At a given location, the linear solution for the inter-laminar stresses is determined. If the stress as a function of radial distance r exceeds the allowable strength from Equation 5.39 or 5.40, respectively, the stress distribution is truncated at the

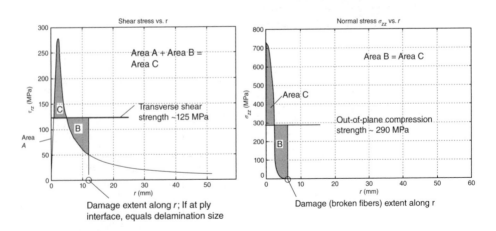

Figure 5.26   Averaged inter-laminar stresses in the damaged region

corresponding allowable strength value. The region over which the stress is equal to the corresponding ultimate strength is obtained by ensuring that the areas under the truncated and non-truncated stress plots are the same, as suggested by the shaded regions in Figure 5.26.

As already mentioned, this is an approximation and does not accurately account for the fact that local damage causes stress redistribution. In a progressive failure evaluation where properties are adjusted locally when damage occurs prior to incrementally increasing the applied load, the final damage size will be somewhat different from the one calculated by this approximate method. In addition, this approach cannot predict if there is unstable growth of the damage. For a more accurate solution, in particular for estimating the size of the delaminations created, one would have to use a fracture mechanics approach. The strain energy release rate would be calculated and compared with its critical value to assess if a given delamination will grow or not. This is not done here. The results will be approximate and, while good in most cases, will not always be valid as will be shown below.

The procedure for determining the type and size of damage during impact is, therefore, as follows:

- Use the approach in Section 5.7.1.1 to calculate the peak force corresponding to a given impact energy level.
- For the given peak force, calculate the stresses σ_{rr}, $\sigma_{\theta\theta}$, σ_z and τ_{rz} at different r, θ and z locations in the vicinity of the impact site.
- If the z location is an interface between plies, the shear stress is decomposed to its τ_{13} and τ_{23} components, each of which is compared to the corresponding allowable ones (1 is parallel to the fibres and may be different for adjacent plies with different orientations). If at least one of the two components exceeds its allowable strength for both plies on either side of the interface in question, there is a delamination. The size of the delamination is found as the distance over which the corresponding stress must be averaged as described in Figure 5.26. If the two plies on either side of the interface in question have the same orientation, then it is assumed that no delamination is possible at that interface.
- The σ_r and σ_θ stresses at the top and bottom of each ply at a given r and θ location are rotated to the ply axis system to obtain stresses parallel and perpendicular to the fibres and the corresponding shear stresses. These are compared with their respective allowable strength to determine if there is failure. This maximum stress criterion has the advantage of determining the type of failure: fibre or matrix, and the mode: tension, compression or shear.
- The previous steps are repeated for different r, θ and z locations until the damage in the entire laminate has been mapped.

Comparison to Test Results

The approach just presented is validated by comparing its predictions to test results. Perhaps the best set of data on quasi-isotropic laminates is the one generated by

Dost *et al.* [12]. In that work, the authors tested a wide variety of quasi-isotropic laminates of the same thickness and showed that their CAI strength at a given energy level could differ by as much as a factor of 1.7 (see next section). A good method to predict damage after impact and, eventually, CAI should be capable of reproducing these results.

The predictions of the present method for three different laminates are compared to the test results in [12] in Figures 5.27–5.29.

To the left of each of the figures is the ultrasonic (C-scan) record of the damage region. This mainly shows the outer envelope of delaminations created at different ply interfaces. Different shades of grey correspond to different through-the-thickness locations but they are not clear enough to allow ply-by-ply differentiation of the predictions. To the right of the same figures are the predictions of the present method. Black lines in the centre correspond to fibre damage at different ply interfaces. The red elliptical lines correspond to out-of-plane normal stress failure at different ply interfaces. Finally, the blue lines in the 'flower-like' shape correspond to delamination sizes at different ply interfaces. Superimposed on the right of each figure is the outer damage envelope from the ultrasonic scan as a continuous purple curve.

It can be seen in all three figures that the damage envelope matches well the predicted shape. And for the different laminates, the overall shape, elongated in Figures 5.27 and 5.28 and more elliptical in Figure 5.29, is captured quite well. The differences between the different lobes predicted by the present method are hard to capture in an ultrasonic scan whose resolution is on the order of a few millimetres,

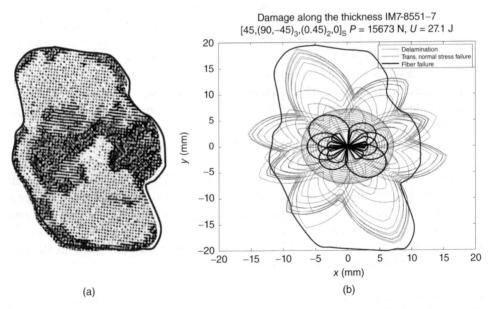

(a)   (b)

Figure 5.27 Ultrasonic scan of damaged region (a) compared to analytical predictions (b) for a [45/(90,−45)$_3$/(0,45)$_2$/0]s IM7/8551-7 laminate with 27.1 J impact (*See insert for colour representation of this figure.*)

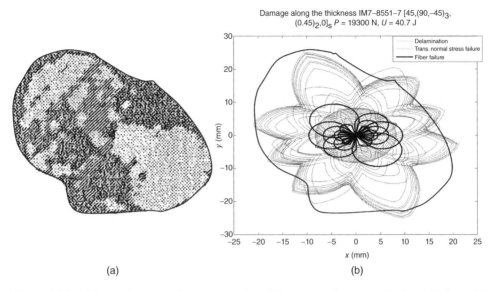

(a)   (b)

Figure 5.28 Ultrasonic scan of damage region (a) compared to analytical predictions (b) for a [45/(90,−45)$_3$/(0,45)$_2$/0]s IM7/8551-7 laminate with 40.7 J impact (*See insert for colour representation of this figure.*)

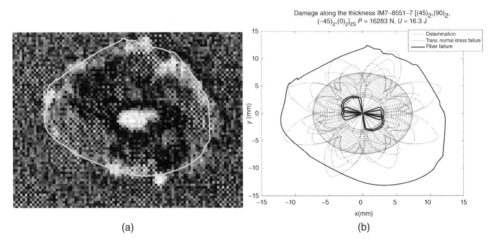

(a)   (b)

Figure 5.29 Ultrasonic scan of damage region (a) compared to analytical predictions (b) for a [45$_2$/90$_2$/−45$_2$/0$_2$]$_{2s}$ IM7/8551-7 laminate with 16.3 J impact (*See insert for colour representation of this figure.*)

which correspond to the differences between the lobes in the right of the three figures. It should be noted that the matrix/fibre damage due to out-of-plane compressive stress failure is not possible to capture by C-scan. Nor is, in general, the fibre failure. However, in some cases, the white spot in the middle may be directly related to fibre damage and seems to correspond qualitatively with the predicted fibre damage shown in Figure 5.29.

The damage shapes in Figures 5.27–5.29 are viewed from the top of the laminate. Through-the-thickness damage maps will be shown in the next section when the CAI predictions are developed.

The comparisons in Figures 5.27–5.29 correspond to specific energy levels. A further comparison can be made by taking the outermost delamination envelope predicted by the present method and comparing it to the corresponding size from the C-scan for different energy levels for various laminates. This is done in Figure 5.30.

Of the seven laminates in Figure 5.30, the first six have 24 plies and the last one has 32 plies. The first and fourth laminates in Figure 5.30 show the largest discrepancy between theoretical prediction and experimental measurements. The remaining five show very good to excellent agreement except for the higher energy levels for laminates 5 and 6. The tests show a plateau there, which probably indicates penetration. For these laminates the impactor started to completely penetrate the laminate, causing a hole. So, further increase of the impact energy does not increase the damage size as it becomes a through-hole.

As the results in Figure 5.30 are all on the same material, another set of test data from Nagelsmit [28] is shown in Figure 5.31. Here the material is AS4/8552 with a very different fibre than the one in Figure 5.30. The resin is similar. Considering the data scatter at low energies, the agreement between predictions and tests is good.

The above results are very encouraging considering the complexity of the problem and the relative simplicity of the method. Ways to improve the method have already been mentioned in connection with most of the assumptions and steps of the approach and will be summarised at the end of this chapter. For now, the approach is quite promising and will be used as the building block for predicting CAI in the next section.

5.7.2   Model for Predicting CAI Strength

The stress analysis in the previous section is used to determine the type and extent of damage caused during impact of a laminate. Depending on the type of damage, reduced local stiffness and strength values are determined creating, effectively, a different laminate locally. Under compressive loading, the local laminate is checked for failure. If there is failure, load is redistributed in surrounding structure and the procedure is repeated until final failure. The laminate locally is assumed to have reduced stiffness and strength properties. This then forms the basis for a CAI analysis.

5.7.2.1   Stress Determination in the Damaged Region

The method is based on the model that treats the damaged region as an inclusion of different stiffness, as described in Section 5.5. This idea was first proposed by Cairns [16] for monolithic laminates and Kassapoglou [29] for sandwiches. Instead of a single inclusion, however, the damaged region is modelled as a series of concentric ellipses, each with its own different stiffness. This allows more accurate evaluation of stresses in the damaged region.

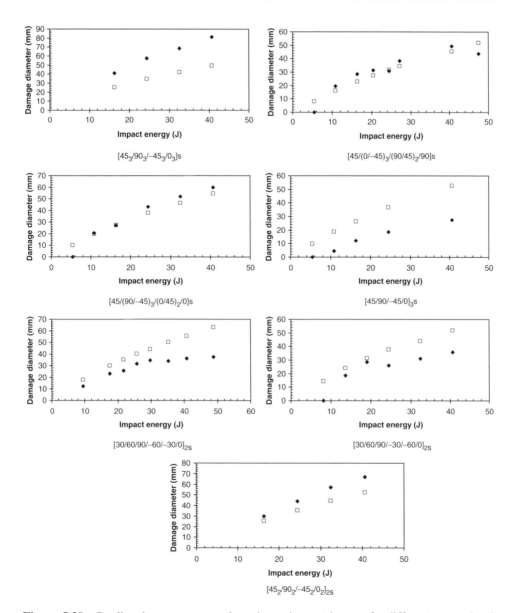

Figure 5.30   Predicted versus measured maximum impact damage for different energy levels for various IM7/8551-7 laminates (open squares: predictions; diamonds: test)

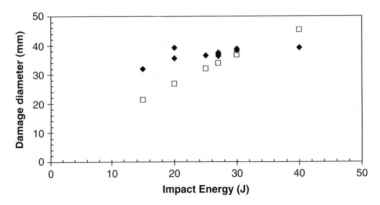

Figure 5.31 Predicted versus measured damage size for $[45/0/-45/90]_{3s}$ AS4/8552 laminate (open squares: predictions; diamonds: test)

The situation is shown in Figure 5.32. The laminate is assumed to be of length $2a$ and width $2b$. A constant compressive displacement u_o is applied at its two ends. The origin of the coordinate system is located at the centre of the damaged region, with x aligned with the loading. The lay-up, including the damaged region, is assumed to be symmetric and balanced.

Double Fourier series are assumed for the displacements u and v in the laminate in the form

$$u = \sum_{m=1}^{\infty} \sum_{n=1}^{\infty} P_{mn} \sin \frac{m\pi x}{a} \cos \frac{n\pi y}{2b} + u_o \frac{x}{a} \tag{5.41}$$

$$v = \sum_{m=1}^{\infty} \sum_{n=1}^{\infty} H_{mn} \cos \frac{m\pi x}{a} \sin \frac{n\pi y}{2b} - \frac{v_{xyeq}u_o y}{a} \tag{5.42}$$

with P_{mn} and H_{mn} unknown constants.

These displacement expressions satisfy the boundary condition that $u = u_o$ at the two ends and the symmetry conditions $u, \partial u/\partial y = 0$ at $x = 0$; and $v, \partial v/\partial x = 0$ at

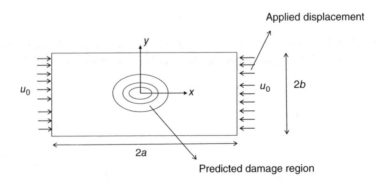

Figure 5.32 Impact damage region modelled as concentric ellipses of different stiffness

$y = 0$. In addition, a term including a weighted-average Poisson's ratio v_{xyeq} is added, given by

$$v_{xyeq} = \frac{\sum_{i=1}^{q} (v_{xy})_i (Area)_i}{4ab} \tag{5.43}$$

where the subscript 'i' denotes the ith elliptical region. It has been found from similar problems that this additional term in Equation 5.42 speeds up convergence of the series involved.

The unknown coefficients P_{mn} and H_{mn} are determined by minimising the energy in the laminate:

$$\Pi_p = U_p - W = \frac{1}{2} \int_0^b \int_0^a \left\{ A_{11} \left(\frac{\partial u}{\partial x} \right)^2 + 2A_{12} \frac{\partial u}{\partial x} \frac{\partial v}{\partial y} \right.$$

$$\left. + A_{22} \left(\frac{\partial v}{\partial y} \right)^2 + A_{66} \left(\frac{\partial u}{\partial y} + \frac{\partial v}{\partial x} \right)^2 \right\} dxdy \tag{5.44}$$

where the work term W is zero because there are no prescribed forces in this problem.

Note that, because of symmetry, only one-quarter of the entire structure in Figure 5.32 needs to be considered. Energy minimization leads to

$$\frac{\partial \Pi_p}{\partial P_{mn}} = 0 \tag{5.45}$$

$$\frac{\partial \Pi_p}{\partial H_{mn}} = 0 \tag{5.46}$$

Evaluating the integrals in Equation 5.44 numerically and applying Equations 5.45 and 5.46 leads to a linear system of $2MN$ equations in $2MN$ unknowns, where M and N are the number of terms in the series in Equations 5.41 and 5.42. Solving the system gives the P_{mn} and H_{mn} values, which can be substituted in the u and v expressions (5.41) and (5.42). Once the displacements are determined, the strains in the laminate are obtained from the strain–displacement equations:

$$\varepsilon_x = \frac{\partial u}{\partial x}$$

$$\varepsilon_y = \frac{\partial v}{\partial y}$$

$$\gamma_{xy} = \frac{\partial u}{\partial y} + \frac{\partial v}{\partial x} \tag{5.47a–c}$$

These, in turn, can be substituted in the constitutive relations

$$\begin{Bmatrix} N_x \\ N_y \\ N_{xy} \end{Bmatrix} = \begin{bmatrix} A_{11} & A_{12} & 0 \\ A_{12} & A_{22} & 0 \\ 0 & 0 & A_{66} \end{bmatrix} \begin{Bmatrix} \varepsilon_x \\ \varepsilon_y \\ \gamma_{xy} \end{Bmatrix} \tag{5.48}$$

to obtain the force resultants. Dividing these by the plate thickness gives the average stresses in the laminate at any location.

Before proceeding, the accuracy of the stresses obtained with this approach must be validated. This was done with two different sets of comparisons. In the first set, the SCF in a quasi-isotropic plate with a circular hole, an elliptical hole with aspect ratio (major/minor axis) of 2, or a circular inclusion with half the stiffness of the surrounding plate was determined. For these cases, exact solutions for infinite plates are available. The SCF is 3 for the circular hole and 2 for the elliptical hole. For the elliptical inclusion with $\lambda = 2$, Equation 5.2 gives a value of 1.49. The laminate used here had lay-up [45/−45/0/90]s with ply thickness 0.152 mm. The length and width of the laminate were equal to 500 mm. The hole diameter was 50 mm. As the theoretical SCFs correspond to infinite plates, a finite width correction was used from Equation 2.5 to correct the theoretical values.

The comparison of the present model to theoretical predictions is shown in Figure 5.33. The present model is within 5% of the theoretical value for the holes and 7% for the elliptical inclusion. The differences are attributed to the fact that the finite width correction factor used is approximate.

The second set of comparisons was done for a more complex case directly related to the problem at hand. Two different laminates, one quasi-isotropic with lay-up [45/−45/0/90]s and one highly orthotropic with lay-up 0_8, each with three concentric ellipses were analysed with FE. The stiffnesses of the three ellipses were, as fractions of the base laminate, equal to 0.2, 0.4 and 0.6, going from the innermost

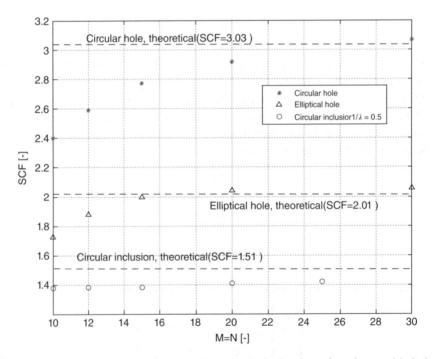

Figure 5.33 Model predictions compared to theoretical values for plates with holes or inclusions

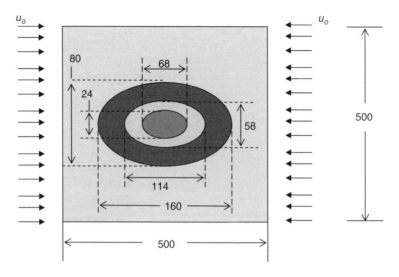

Figure 5.34 Laminate with three elliptical inclusions (dimensions are in millimetres)

to the outermost ellipse. The material used was AS4/8552. The analysis was done with ABAQUS using reduced integration shell elements. The situation is shown in Figure 5.34.

The comparison of the present model to the FE results for the quasi-isotropic laminate is shown in Figure 5.35. The corresponding results for the 0_8 orthotropic laminate are shown in Figure 5.36. The axial stresses in the direction of the load normalised by the far-field stress are plotted along a transverse line (parallel to y in Figure 5.32) starting from the centre of the innermost ellipse and moving outboard.

In both cases very good agreement between the present model and FE is observed. Even for the highly orthotropic 0_8 laminate, where the mismatch in stiffnesses is expected to cause difficulties, the agreement is very good. Only at the very centre of the 0_8 laminate there is a discrepancy. There is also a strange shape of the curves inside each of the three elliptical inclusions where the present model predicts convex curve shapes while the FE results are concave. The differences, however, are very small and the discontinuities from one inclusion to the next are captured accurately. The present model is considered accurate enough to use for predicting CAI strength.

5.7.2.2 Model for Predicting CAI Strength

In order to use the model from the previous section, the laminate, damaged as a result of impact, must be divided into elliptical regions of known reduced stiffness. This is done by resorting to the damage maps created using the method of Section 5.7.1.3. Here, the through-the-thickness maps, as shown in Figure 5.37, are more useful.

At any θ location a map similar to the one shown in Figure 5.37 can be created. As is seen from Figure 5.37, the damage is not symmetric with respect to the mid-plane

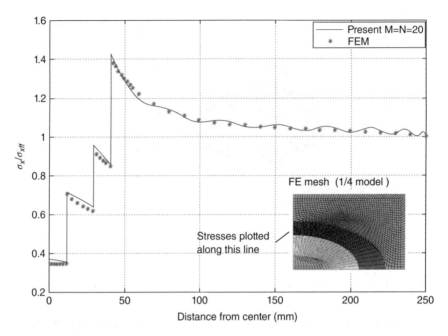

Figure 5.35   Stresses in a laminate with elliptical inclusions. Model predictions versus FE for a [45/−45/0/90]s laminate

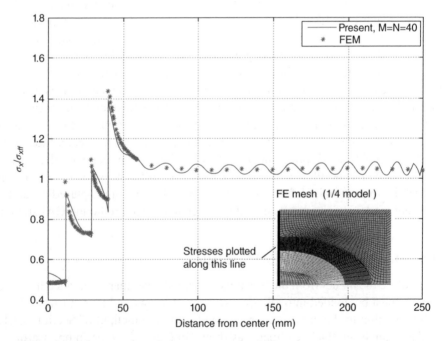

Figure 5.36   Stresses in a laminate with elliptical inclusions. Model predictions versus FE for a 0_8 laminate

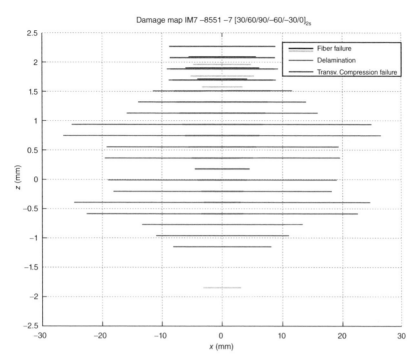

Figure 5.37 Extent and type of damage at $\theta = 0$ for a $[30/60/90/{-}60/{-}30/0]_2$s IM7/8551-7 laminate (*See insert for colour representation of this figure.*)

of the laminate. Also, the pattern will change with θ location. Note that matrix failure is not shown in this figure to avoid cluttering it up. As a first approximation, the division in regions is done by separating the region that has fibre failure from that which does not. The centre of the laminate extending from approximately -7 to $+7$ mm becomes the innermost region. Then, the region extending to the largest delamination, at ± 28 mm approximately, is the second region. Outboard of this region, the laminate is assumed to be intact even though some matrix damage may be present. These dimensions correspond to one θ location. By going around from $0°$ to $180°$, the biggest radial distances over which there is fibre damage and delamination can be determined. These greatest distances form the boundaries of the regions. The overall sizes across all θ values are shown in Figure 5.38 (energy level is lower than that in Figure 5.37). The subdivision region is also shown in Figure 5.38.

For each of the regions in Figure 5.38, updated stiffness properties are computed. This is a crucial step. The results will depend on how accurately the stiffness and strength degradation are simulated in the model. There are several ways of doing this: from the conservative approach where the properties affected by damage in a ply are set to zero, to more realistic approaches where the properties are updated on the basis of a continuum damage model [30]. It is recognised here that setting properties in a ply to (near) zero values is not realistic. Stresses in a damaged ply can be transferred

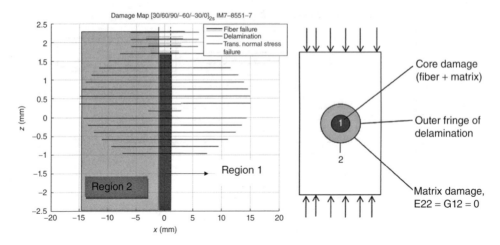

Figure 5.38   Division of damaged laminate into regions (*See insert for colour representation of this figure.*)

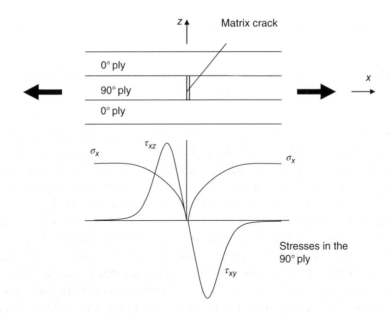

Figure 5.39   Stress transfer around a matrix crack in a 0/90/0 laminate under tension

around a damaged region into adjacent undamaged plies and then can be transferred back again in undamaged regions of the damaged ply. A simple example is shown in Figure 5.39 for a 0/90/0 laminate under tension (0 direction parallel to x).

Under tension, the 90° ply will develop matrix cracks perpendicular to the applied load. One of these cracks is shown in Figure 5.39. As long as the adjacent matrix cracks are sufficiently far from this one, the stresses in the 90° ply will, qualitatively, have the shape shown at the bottom of the figure: The axial stress σ_x will drop to zero at the crack location. The axial load in the 90° ply will be transferred to the adjacent 0° ply through the shear stress τ_{xz}, which develops in the vicinity of the crack, reaches a maximum and drops again to zero at the crack surface. If the 90° ply did not have the 0° plies on either side, as soon as the first crack developed, the laminate would fail completely. This is not the case in Figure 5.39, where significantly more load beyond the load at which the first transverse cracks appears can be reached before the 0° plies fail for complete failure of the laminate. This example was analyzed in detail in Section 3.5 and will be revisited in more detail in Section 6.6.2.

The example in Figure 5.39 is oversimplified to illustrate the point. In a laminate under compression, similar mechanisms may be present, effectively allowing damaged plies to carry load beyond onset of failure. In fact, even for a single angle ply under tension, (reduced) load transfer is possible beyond first failure as fibres pull out of the matrix and longitudinal splits develop. This has been studied extensively by several researchers. The most comprehensive study can be found in [30].

Knowledge of how load is transferred in the vicinity of damage requires extensive testing and simulation of the failure process, but there is no single definitive continuum damage model that can be used in all cases. Here, two different models will be used, each with different amount of load transfer capability assigned to the damaged plies after first failure. This will provide a range of answers and will give an indication of how sensitive the final results can be to the different models.

Model 1 (Load Transfer through Damaged Plies Allowed)
In each of the two inboard regions in Figure 5.38, the properties in a damaged ply are updated as follows:

- If there is fibre failure which extends from the top to the bottom of the ply, the ply strength values along the fibres X^t and X^c are reduced to 20% of their undamaged values. The corresponding stiffness E_{11} is reduced to 5% of its undamaged value.
- If there is fibre failure but does not extend from the top to the bottom of the ply, X^t and X^c are reduced to 50% of their undamaged values and E_{11} to 5% of its undamaged value.
- If there is matrix failure anywhere in the ply, the matrix-dependent properties Y^t, Y^c, S, E_{22} and G_{12} are reduced to 5% of their undamaged values.
- The above updates are effective over the radial distance from the centre of impact over which the corresponding failures occurred.

Model 2 (No Load Transfer Allowed through Damaged Plies)
In each of the two inboard regions in Figure 5.38, the properties in a damaged ply are updated as follows:

- For fibre failure, E_{11}, X^t, X^c are reduced to 1% of their undamaged value.
- For matrix failure, E_{22}, G_{12}, Y^t, Y^c and S are reduced to 5% of their undamaged values.

- The above updates are effective over the radial distance from the centre of impact over which the corresponding failures occurred.

Once a model has been selected and the ply-level stiffness values have been updated, the new A matrix (and D matrix) for each sub-laminate can be calculated. This is then used to determine local strains and stresses. A maximum stress failure criterion with the updated strength values is used to check if there is further failure or not.

If now the innermost region in Figure 5.38 fails, the load carried by it is redistributed to the region surrounding it and the undamaged laminate around it using strain compatibility (all laminates strain by the same amount). The damaged laminate now behaves as if there is a hole in the middle. If there is no failure, the load is increased until the second sub-laminate fails and the load is redistributed to the surrounding undamaged laminate. The load is increased until final failure.

If, instead of the inner region, the outer region fails first, the laminate is modelled as having a hole of size equal to the two regions. Load is again increased until final failure.

5.7.2.3   Comparison to Test Results

The approach presented in the previous section was applied to the test results from [12]. For convenience, these test results are repeated in Figure 5.40. As already mentioned, even though the laminates in Figure 5.40 all have the same thickness and same

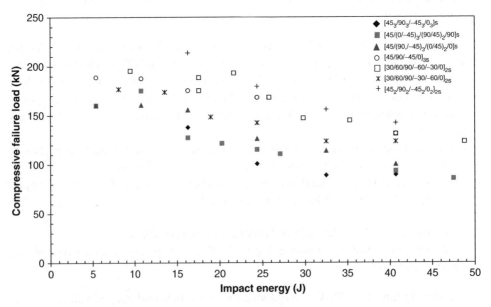

Figure 5.40   Compression after impact (CAI) failure loads for quasi-isotropic laminates (*Source*: From Dost *et al.* [12].) (*See insert for colour representation of this figure.*)

in-plane stiffness in every direction, they have drastically different CAI strengths, with the highest differing from the lowest by as much as a factor of 1.7 for some impact energies.

Predictions of the present method using Model 1 from the previous section are compared to test results in Figure 5.41. The corresponding comparison using Model 2 for the predictions is shown in Figure 5.42.

Comparing the two figures, it can be seen that only for laminates *b* and *c* does the model used make a significant difference. In all other cases, the differences between the two models are small. Interestingly, for laminates *b* and *c*, it appears that use of Model 2 gives much better results with only small differences between analysis and tests. However, this is not necessarily sufficient reason to prefer Model 2 over Model 1. Neither model allows for sub-laminate buckling. Sub-laminate buckling has not been incorporated in the analysis, but it is known to be the primary failure mode for some laminates with damage sizes greater than 50 mm. This makes sense if one considers that these damage sizes mostly represent the extent of delaminations. And for such sizeable delaminations, sub-laminate buckling, for example using the approach in Section 4.3.2, should also be examined. With this in mind, Model 1 might be the preferred one provided that the discrepancies from test results for laminates b, c and g can be attributed to sub-laminate buckling. If sub-laminate buckling is neglected, Model 2 appears to be the most accurate with only one of seven laminates, laminate g, showing significant differences between analysis and test.

5.7.2.4 Discussion

The results presented in the previous section are encouraging. They suggest that a stress-based model can be used to get a good estimate of the extent and type of damage during impact of a quasi-isotropic laminate. Then, by modelling the damaged region as elliptical inclusions of different stiffness consistent with the local damage, one can obtain a good estimate of the CAI strength of the laminate.

This approach, while very promising, has several drawbacks and areas where more work is needed. Some of the issues were mentioned in passing in the two previous sections. Here they are summarised as areas of on-going and future research.

- Sub-laminate buckling must be included in the failure analysis. In particular the interaction between material failure and sub-laminate buckling is expected to be a key factor in improving the predictions of the current model. One of the problems with sub-laminate buckling is determining the exact shape of the delaminated sub-laminates. A conservative approach using the model of Section 4.3.2 and elliptical sub-laminates showed buckling occurring too early. An accurate model for sub-laminate buckling and a method to partially transfer load from buckled sub-laminates to unbuckled ones are needed.
- The failure criterion used here was a maximum stress criterion. Improved criteria such as the Puck criterion [31] are expected to give more accurate results. In the

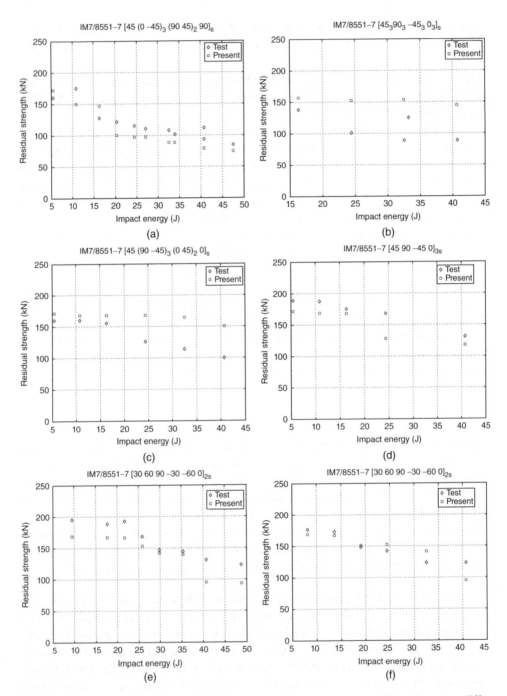

Figure 5.41 (a–g) Analytical predictions versus test results for CAI failure for seven different laminates (damaged plies carry load according to Model 1)

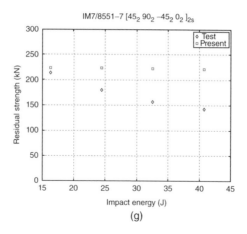

Figure 5.41 (*Continued*)

case where the inclusions fail completely and the laminate behaves as if there is a hole in the middle, an improved failure model for laminates with elliptical holes is needed.

- The two models used for progressive failure must be improved on the basis of better modelling of post-damage behaviour of the laminate. While the results changed little for most laminates when the failure model was changed, the difference for a couple of laminates is too significant to ignore. Methods such as the one by Maimí *et al.* [30] are good candidates for implementation instead of the two models proposed here.

- The subdivision of the damaged laminate in two inclusions proposed here was somewhat arbitrary by choosing fibre damage and delaminations as the guidelines for the inclusion boundaries. More accurate results can be obtained if more inclusions are used. This will increase the computation time but may be worthwhile for some laminates.

- As long as the driving failure mode is material failure, as opposed to sub-laminate buckling, the actual size and shape of the damaged region has a small effect on the final predictions. The results presented above were generated using circular inclusions. It was found that changing them from circular to elliptical with minor axis 90% of the length of the major axis changed the predictions by less than 1% for five laminates and 15% for the other two. Also, changing the overall size had a very small effect. Only if the size is such that finite width effects are significant would the size (and shape) become more of a driver.

- All the laminates used in the above comparison were quasi-isotropic. While the approach to determine the type and extent of damage in Section 5.7.1 is restricted to quasi-isotropic laminates, the approach in Section 5.7.2 for predicting CAI loads is not. As long as the damage footprint can be represented well by an ellipse, the method will still be valid. Only for very highly orthotropic laminates (e.g. all 0° laminates) where the longitudinal splits extent over large distances and form an irregular rectangle would the present approach not be applicable.

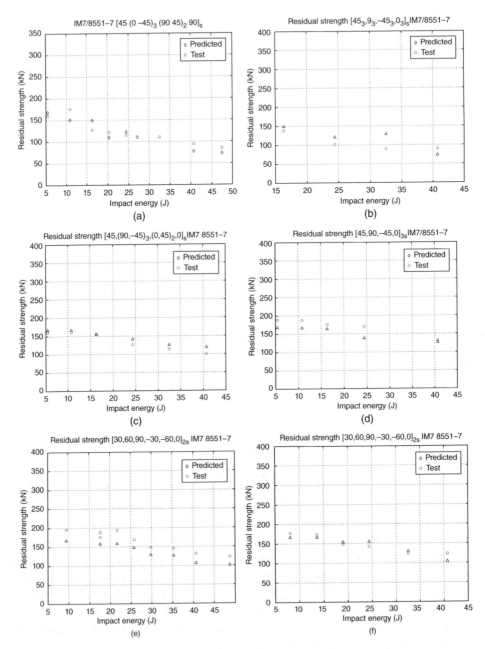

Figure 5.42   (a–g) Analytical predictions versus test results for CAI failure load for seven different laminates (damaged plies carry no load according to Model 2)

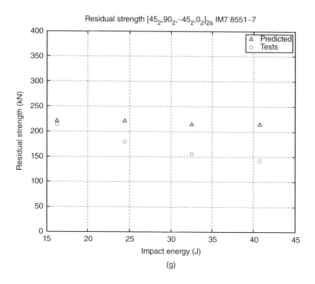

Figure 5.42   (*Continued*)

Exercises

5.1   For a single ply of any orientation under uniaxial tension N_x, derive an expression for its effective axial stiffness E_x assuming that, during the uniaxial tension test, $N_y = N_{xy} = 0$.

5.2   A material with the following parameters is used in a CAI test.

$$
\begin{aligned}
E_x \text{ tension} &= 137.8\,\text{GPa} \\
E_x \text{ compr} &= 114.8\,\text{GPa} \\
E_y \text{ tension} &= 8.58\,\text{GPa} \\
E_y \text{ compr} &= 8.95\,\text{GPa} \\
\nu_{xy} &= 0.29 \\
G_{xy} &= 4.92\,\text{GPa} \\
t_{\text{ply}} &= 0.188\,\text{mm} \\
X^t &= 2042\,\text{MPa} \\
X^c &= 1495\,\text{MPa} \\
Y^t &= 66.1\,\text{MPa} \\
Y^c &= 257\,\text{MPa} \\
S &= 105.2\,\text{MPa}
\end{aligned}
$$

The laminate $[(45/{-}45/0/90)_3]$s is impacted at BVID. C-scan shows circular damage with diameter 38 mm. Measurements have shown that the energy to cause BVID is 25 J and the peak force during impact is 9000 N.

(a) Determine the deflection δ at maximum force.
(b) Assume that, during impact, the laminate conforms to the shape of the steel impactor of radius $R = 16\,mm$ as shown below, and determine the contact radius r_c at peak load.

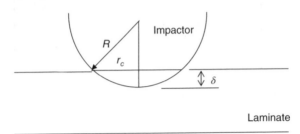

(c) Divide the impact region into three sub-regions: the one in the middle of dimensions $2r_c$ and the two outer portions. Assume that the portion below $2r_c$ is completely failed (can you provide a justification for this assumption?). Also assume that the two outer portions consist of completely delaminated plies (i.e. there is one delamination at every ply interface in the two outer regions). By doing a one-dimensional analysis along the 0 direction, determine the stiffness of the damaged region (you will need the result of Exercise 5.1 here).
(d) Estimate the CAI strength in two ways: first, using analysis to predict both the undamaged and the damaged compression strength, and, second, using the test result of 450 MPa for the undamaged compression strength. How do your predictions compare to the test result (with BVID) of 200 MPa?

5.3   A sandwich structure has quasi-isotropic facesheets. When impacted, the damage region is circular of radius R_o. Depending on the impact energy, the damage ranges from low, where the centre of the damage region has just a tiny pinhole, to high, where a hole of radius R_i is created. Assume that the stiffness inside the damage region varies linearly from zero at the edge of the inner hole caused by impact to E_f, the far-field (undamaged) stiffness. Also assume that, for purposes of estimating the effect of impact damage, the overall stiffness E_2 of the damaged region can be estimated as the average stiffness over the entire damage region (region of radius R_o) (Figure E5.1).
(a) Derive an expression for the SCF as a function of R_i/R_o as R_i ranges from 0 to R_o.
(b) Show that in the two cases: (i) no damage and (ii) through hole (i.e. the impactor creates a hole of radius R_o), your expression in (a) recovers well-known results.
(c) Plot SCF versus R_i/R_o.

(d) As mentioned in Section 5.1, BVID is defined in terms of the indentation depth, which, among other things, puts serious strain on the eyesight of an inspector. Assume that BVID for a sandwich corresponds to an impact energy that causes just enough damage to barely start creating a tiny hole at the centre of the impact site. If the failure stress for the facesheet (without damage) is σ_{ult}, what is the failure stress with BVID?

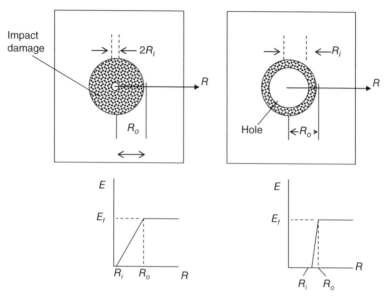

Figure E5.1   Postulated stiffness variation as a function of distance in impacted sandwich

5.4   (a) The properties of the AS4/8552 unidirectional graphite epoxy material are as follows:

$$E_x \text{ tension} = 137.8 \text{ GPa}$$
$$E_x \text{ compr} = 114.8 \text{ GPa}$$
$$E_y \text{ tension} = 8.58 \text{ GPa}$$
$$E_y \text{ compr} = 8.95 \text{ GPa}$$
$$v_{xy} = 0.29$$
$$G_{xy} = 4.92 \text{ GPa}$$
$$t_{ply} = 0.188 \text{ mm}$$
$$X^t = 2042 \text{ MPa}$$
$$X^c = 1495 \text{ MPa}$$
$$Y^t = 66.1 \text{ MPa}$$
$$Y^c = 257 \text{ MPa}$$
$$S = 105.2 \text{ MPa}$$

The following laminate

$$[(45/-45/0/90)_3]s$$

is tested in compression (specimen is 180 mm long and 127 mm wide with applied load along the long dimension) and its strength is found to be 401.8 MPa. It is then impacted with 25 J of impact energy. The resulting damage as measured by delaminations is shown in the table below. The CAI is 191 MPa. Concentrating ONLY on the delaminations, obtain a prediction for the CAI strength. Use the delamination information from the following table. Note ply number 1 is the top ply (Tables E5.1 and E5.2).

Table E5.1 Delaminations starting from centre and extending to the left

Delamination between plies	Length (mm)
2/3	4.31
4/5	6.55
6/7	8.45
8/9	6.21
10/11	9.31
11/12	1.72
13/14	12.76
15/16	8.62
17/18	12.76
20/21	10.52
22/23	8.62

Table E5.2 Delaminations starting from centre and extending to the right

Delamination between plies	Length (mm)
2/3	0.69
6/7	7.76
8/9	6.03
11/12	1.38
13/14	14.66
17/18	11.90
22/23	14.31

(b) Compare your prediction to the test result. Comment on reasons for any differences.

5.5 You are investigating engineering detectives in search for the elusive 'best laminate' for CAI. You are looking for symmetric and balanced laminates. After 'canvassing' the possibilities and collecting all kinds of information, you decide that the best laminate would be one that, given the undamaged to damaged stiffness ratio λ, minimises the corresponding SCF at the edge of the damaged region. You then use the expression given in Section 5.5 for the SCF and you notice it is a function of three parameters that refer to the undamaged laminate, E_{22}/E_{11}, E_{11}/G_{12} and v_{12}. You start varying these values and you notice that the resulting SCF goes through a minimum under certain circumstances. However, you cannot vary the three parameters independently of one another and this complicates (and also simplifies) things. A friend from TUDelft reminds you that you can express every in-plane property in terms of the lamination parameters V_1 and V_3, thereby reducing the three unknowns to two, and incorporating the relative dependence of the different parameters more easily (i.e. the limits on V_1 and V_3 depend on each other). Your good friend also gives you the following relations:

$$V_3 = \frac{U_2{}^2 V_1{}^2 - U_2 E_{11} V_1 + E_{11} U_1 - U_1{}^2 + U_4{}^2}{U_3(2U_1 + 2U_4 - E_{11})}$$

$$= \frac{U_2{}^2 V_1{}^2 + U_2 E_{22} V_1 + E_{22} U_1 - U_1{}^2 + U_4{}^2}{U_3(2U_1 + 2U_4 - E_{22})}$$

$$= \frac{v_{12} U_2 V_1 - v_{12} U_1 + U4}{(1 + v_{12})U_3} = \frac{U_5 - G_{12}}{U_3}$$

where U_i are standard laminate invariants and E_{11}, E_{22}, G_{12} and v_{12} are laminate quantities.

You also notice that

$$-1 \leq V_1 \leq 1$$

$$V_3 \geq 2V_1^2 - 1$$

for a general laminate.

You can now solve the problem.

(a) Determine the symmetric and balanced laminate consisting of ONLY 0, 45, −45 and 90 plies that minimises the SCF for $\lambda = 1.4$, 2, 3 and 5. Note that in this case the ranges on V_1 and V_3 change. This range of λ values should definitely cover BVID energies for most laminates. Why does your answer make perfect sense in the limiting case where $\lambda \to \infty$? (Discuss this by considering test data presented in class). Also, why is the 'winning laminate' you came up with a rather bad (in fact terrible) laminate for good CAI strength?

(b) Determine the V_1 and V_3 values that correspond to the best symmetric and balanced laminate satisfying the 10% rule for the same λ values as in part (a). Create a plot of minimum SCF versus λ. (Bonus question: Try to determine the best lay-up for a 16-ply laminate with $\lambda = 3$.)

(c) Discuss when your analysis above breaks down.

References

[1] Kassapoglou, C. (2013) *Design and Analysis of Composite Structures*, 2nd edn, Chapter 5.1.5, John Wiley & Sons, Inc., New York.

[2] Lagacé, P.A. (1986) Delamination in composites: is toughness the key? *SAMPE J.*, **22**, 53–60.

[3] NASA (1983) *Standard Test for Toughened Resin Composites*, NASA Reference Publication 1092, Langley Research Center, Hampton, VA.

[4] Avery, J.G., Porter, T. and Walter, R.W. (1972) Designing aircraft structure for resistance and tolerance to battle damage. AIAA 4th Aircraft Design, Flight Test, and Operations Meeting, LA, CA, 1972, AIAA-1972-773.

[5] Williams, J.C. (1984) Effect of Impact Damage and Open Holes on the Compression Strength of Tough Resin/High Strength Fiber Laminates. NASA-TM-85756.

[6] Puhui, C., Zhen, S. and Junyang, W. (2002) A new method for compression after impact strength prediction of composite laminates. *J. Compos. Mater.*, **36**, 589–610.

[7] Lekhnitskii, S.G. (1963) in *Theory of Elasticity of an Anisotropic Elastic Body* (translated by P. Fern), Holden Day Inc., San Francisco, CA.

[8] Savin, G.N. (1961) in *Stress Concentration Around Holes* (translated by W. Johnson), Pergamon Press.

[9] Kassapoglou, C., Jonas, P.J. and Abbott, R. (1988) Compressive strength of composite sandwich panels after impact damage: an experimental and analytical study. *J. Compos. Tech. Res.*, **10**, 65–73.

[10] Nyman, T., Bredberg, A. and Schoen, J. (2000) Equivalent damage and residual strength of impact damaged composite structures. *J. Reinf. Plast. Compos.*, **19**, 428–448.

[11] Kassapoglou, C. (2013) *Design and Analysis of Compostie Structures*, 2nd edn, Chapter 10, John Wiley & Sons, Inc., New York.

[12] Dost, E.F., Ilcewicz, L.B., Avery, W.B. and Coxon, B.R. ASTM STP 1110 (1991) Effect of Stacking Sequence on Impact Damage Resistance and Residual Strength For Quasi-Isotropic Laminates, ASTM, pp. 476–500.

[13] Lekhnitskii, S.G. (1968) *Anisotropic Plates*, Chapter VI-43, Gordon and Breach Science Publishers, New York.

[14] Esrail, F. and Kassapoglou, C. (2014) An efficient approach for damage quantification in quasi-isotropic composite laminates under low speed impact. *Composites Part B*, **61**, 116–126.

[15] Esrail, F. and Kassapoglou, C. (2014) An efficient approach to determine compression after impact strength of quasi-isotropic composite laminates. *Compos. Sci. Technol.*, **98**, 28–35.

[16] Cairns, D.S. (1987) Impact and post-impact response of graphite/epoxy and kevlar/epoxy structures. PhD thesis, Department of Aeronautics and Astronautics, Massachusetts Institute of Technology.

[17] Olsson, R. (2001) Analytical prediction of large mass impact damage in composite laminates. *Composites Part A*, **32** (9), 1207–1215.

[18] Kassapoglou, C. (2014) *Design and Analysis of Composite Structures*, Chapter 5.3.2, 2nd edn, John Wiley & Sons, Inc., New York.

[19] Hertz, H. (1895) *Gesammelte Werke*, vol. **1**, J.A. Barth, Leipzig.

[20] Lesser, A.J. and Filippov, A.G. (1994) Mechanisms governing the damage resistance of laminated composites subjected to low-velocity impacts. *J. Reinf. Plast. Compos.*, **3**, 408–432.

[21] Shivakumar, K.N., Elber, W. and Illg, W. (1983) Prediction of Impact Force and Duration During Low Velocity Impact on Circular Composite Laminates. NASA TM 85703.

[22] Talagani, F. (2014) Impact analysis of composite structures. PhD thesis. Delft University of Technology, Delft.

[23] Christoforou, A.P. and Yigit, A.S. (1995) Transient response of a composite beam subjected to elasto-plastic impact. *Compos. Eng.*, **5**, 459–470.

[24] Yang, S.H. and Sun, C.T. (1982) Indentation law for composite materials, in *Composite Materials: Testing and Design (6th Conference)* ASTM STP 787 (ed I.M. Daniel), pp. 425–449.

[25] Kassapoglou, C. and Lagacé, P.A. (1986) An efficient method for the calculation of interlaminar stresses in composite materials. *J. Appl. Mech.*, **53**, 744–750.

[26] Kassapoglou, C. (1990) Determination of interlaminar stresses in composite laminates under combined loads. *J. Reinf. Plast. Compos.*, **9**, 33–59.

[27] Love, A.E.H. (1929) The stress produced in a semi-infinite solid by pressure on part of the boundary. *Philos. Trans. R. Soc. London*, **228**, 377–420.

[28] Nagelsmit, M. (2013) Fiber placement architectures for improved damage tolerance. PhD thesis. Delft University of Technology.

[29] Kassapoglou, C. (1996) Compression strength of composite sandwich structures after barely visible impact damage. *J. Compos. Tech. Res.*, **18**, 274–284.

[30] Maimí, P., Camanho, P.P., Mayugo, J.A. and Dávila, C.G. (2007) A continuum damage model for composite laminates – part I: constitutive model. *Mech. Mater.*, **39**, 897–908.

[31] Puck, A. and Schürmann, H. (1998) Failure analysis of FRP laminates by means of physically based phenomenological models. *Compos. Sci. Technol.*, **58**, 1045–1067.

6

Fatigue Life of Composite Structures: Analytical Models

6.1 Introduction

Under repeated loading, even if the load magnitudes are significantly lower than the corresponding static strength of the structure, the stiffness and the strength degrade. After a sufficient number of cycles, the structure fails. Fatigue is a process via which the load-carrying ability of a composite structure diminishes as a function of cyclic loading. This happens because some type of damage is created and grows reducing both the stiffness and the strength. Damage creation and evolution in composites occur at different scales starting, for example, from very low scales where tiny voids coalesce into microcracks or matrix crazing occurs. As cycling continues, damage grows or triggers other types of damage, for example, matrix cracks may transition to delaminations. At some point, the extent and the type of damage are such that the structure cannot sustain any further cyclic loading and failure occurs. Usually, the final failure is associated with fibre breakage but is not uncommon to have structure failing because the number and the size of delaminations have reduced the bending stiffness to a point that the deflections are excessive and the structure can no longer perform its function.

Because damage may start at very low-length scales, in the order of a fibre diameter or less, it may go undetected for a long time. Only after it has evolved or grown to a damage type and size measurable by the inspection means used, it is possible to detect and monitor it. As mentioned in Chapter 1 and discussed in some detail in Chapters 2–5, three generic types of damage are distinguished in composite structures: (i) matrix cracks, (ii) delaminations and (iii) fibre breakage. Variations of these, such as longitudinal splits, transverse cracks, fibre kinking and combinations such as delaminations emanating from matrix cracks terminating at ply interfaces, are quite common.

In typical applications of composite materials, damage representative of what may occur during service does not grow at all for a relatively large number of cycles.

Modeling the Effect of Damage in Composite Structures: Simplified Approaches, First Edition. Christos Kassapoglou.
© 2015 John Wiley & Sons, Ltd. Published 2015 by John Wiley & Sons, Ltd.

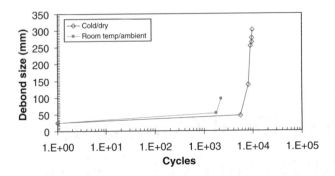

Figure 6.1 Growth of delamination in sandwich with (±45) fabric facesheets. Shear loading

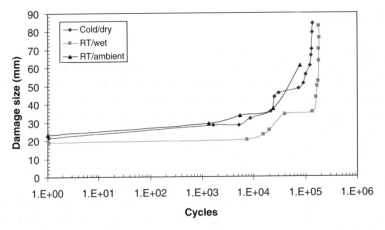

Figure 6.2 Damage growth (mostly delaminations) from a cross-head crack in sandwich with 42% 0, 46% ±45, 12% 90 tape and fabric facesheets. Tension/compression loading

Once growth starts, it is quite fast. Typical test data are shown in Figures 6.1 and 6.2. In Figure 6.1, the test data were obtained on 500 mm × 500 mm square sandwich panels under shear. In Figure 6.2, the test data were obtained on 150 mm × 300 mm sandwich panels under tension/compression loading.

As can be seen from Figure 6.1, there is no measurable growth for 6000 cycles (Cold/Dry condition) and then rapid growth to failure for 4000 cycles. Very similar behaviour is observed in Figure 6.2 for the Cold/Dry and Room Temperature/Wet conditions where almost no growth is observed for 20,000 cycles, then a relatively slow growth for about 70,000 cycles and, finally, a rapid growth for the last 100,000 cycles. It also appears that there is an environmental effect: Room Temperature Ambient specimens perform worse, that is, show faster growth than other environmental conditions at least for the material, layups and loading covered in Figures 6.1 and 6.2. Another important observation is that growth, when it finally starts, is not smooth. It occurs in jumps as can be seen in the region between 10,000 and 80,000 cycles for Cold/Dry and RT/Wet in Figure 6.2.

This type of behaviour occurs often in typical composite aircraft structures. Growth is relatively rapid and non-uniform. Also, several types of damage may be combined. In Figure 6.2, damage growth consisted mostly of delaminations emanating from the cracks but, towards the end of the test, some crack growth, which was not self-similar, was also observed. Cases of 'well-behaved' relatively slow growth of a single type of damage, usually a delamination, do occur in practice but are not as common. In addition, the experimental scatter associated with damage growth at these sizes of specimens and damage, is quite large. Detection can be difficult and inspection records may have difficulty differentiating between damage types and their location. For these reasons, establishing economically feasible inspection intervals for composite structures on the basis of damage growth is complicated and expensive.

Despite these issues, composites have very good fatigue behaviour because of their relatively flat S–N curves. The term *S–N curve* is used here somewhat loosely. In a metal, the S–N curve defines the stress level needed to cause a specific type of failure after a certain number of cycles. Depending on how it is defined, it consists of up to two relatively well-defined regimes: crack nucleation/initiation and crack propagation. The key points are that there is one failure mechanism that of unstable crack growth and a single, mostly self-similar in its evolution, type of damage, the crack. In composites, as already mentioned, one must keep in mind that, even during constant amplitude loading, multiple types of damage may be present that may interact switching from one type dominating to another and creating a damage pattern that is rarely self-similar. Therefore, a single S–N curve describing cycles to failure for a specific composite structure and loading implies that, at different points along the curve, the failure mode may be different and the damage pattern leading to failure may be different. This is important because if there were a single failure mode, one could 'exchange' cycles with loads. Increasing the load amplitude would lead to the same type of failure but earlier. Doing the same thing in composites leads to shorter lives (as a rule) but not necessarily in the same failure mode. This has important implications for effects of load sequencing, which will be discussed later.

A 'thought experiment' that demonstrates how the different failure modes may interact in counter-intuitive ways follows. In fact, it will be suggested that if the conditions of the 'experiment' are met, specimens with different static strength may have reverse order in residual strength after fatigue loading. That is, if one specimen has higher static strength, it may end up having lower residual strength after cycling.

Consider two nominally identical specimens A and B as shown in Figure 6.3. They consist of only unidirectional tape material aligned with the loading direction and they have a hole at their centre. The specimen dimensions are large enough so there are no finite-width effects.

Assume now that, due to the manufacturing process, tiny flaws may be present in the specimens that have escaped inspection so both specimens are considered adequate for testing. However, in specimen A, the tiny flaw is close to the hole edge while in

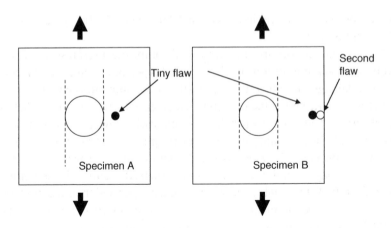

Figure 6.3   Nominally identical specimens under tension–tension fatigue loading

specimen B it is far away from the hole edge (see Figure 6.3). In a static test, specimen A will fail at a lower load because the flaw is well within the region of high stresses due to the stress concentration effect of the hole. It will therefore trigger the failure of specimen A earlier than the flaw in specimen B. This is in line with the discussion in Section 2.4 in relation to Figure 2.10.

Suppose now that both specimens are loaded in tension–tension fatigue ($R = \sigma_{min}/\sigma_{max} = 0$). And that the applied maximum cyclic load is relatively low so the specimens do not fail after a few cycles. It is well known that longitudinal splits will develop in these specimens emanating from the hole edges and extending up and down (parallel to the loading). These splits are shown as dashed lines in Figure 6.3. They do not necessarily extend all the way through the laminate thickness so they do not physically split the specimen into strips. Their effect, however, is to render the centre strip above and below the hole ineffective. The applied load, therefore, is transmitted through the two outer strips in each specimen eliminating the stress concentration effect of the hole. It is also known that once these splits develop, if the fatigue test is stopped and a static test to failure is performed, the residual strength of these specimens will be higher than their static strength because the effect of the hole has been, effectively, eliminated. This means that specimens A and B will now have equivalent residual strength, even though they started with significantly different static strength values with specimen B the stronger of the two.

It is now relatively easy to take one additional step in the thought process. Suppose specimen B has another non-detectable flaw next to the first one, designated by the open circle in Figure 6.3. Now, specimen B will have a lower residual strength after cycling compared to specimen A because of the combined effect of the two flaws.

The above is a hypothetical situation but is not impossible. And, in the limiting case, where the applied stress equals the residual strength, when the cycles to failure have been reached, it serves as a counter-example to the 'equal rank assumption' [1]. According to the equal rank assumption, specimens of the same population will maintain the same rank during fatigue cycling as they do for static strength. That is, S–N curves starting at two different static strength values before cycling do not cross for any number of cycles. It should be emphasised that there is no evidence that the 'equal rank assumption' is not valid in most cases. The previous 'thought experiment' was not meant to invalidate the 'equal rank assumption'. Its purpose is to underline the fact that extra care must be exercised in interpreting test results and establishing analytical models.

The preceding discussion was meant to highlight some of the complexities of fatigue of composites. It is not a detailed discussion. Excellent detailed overviews of the topic can be found in specialised references [2–4]. The purpose of this chapter is to provide some approaches that may be useful in developing analytical models for predicting fatigue lives of composite structures.

Given the increased complexity of damage creation and evolution during fatigue, creating an analytical model to predict cycles to failure, even for constant amplitude loading, can be very challenging. Typically, for higher accuracy, some form of curve-fitting to test data is necessary. And if curve fitting is avoided, either the model has limited applicability (for example, only for a class of layups or R values) or, it is computationally very expensive as it attempts to characterise and track all possible forms of damage from one cycle to the next.

6.2 Needed Characteristics for an Analytical Model

The emphasis in this chapter is on setting up analytical models that are not just curve fits and do not rely on test results to fix model parameters. This being quite a challenge, first some basic model requirements and desired characteristics will be discussed.

Conceptually, an analytical model must be able to determine when damage starts and how it grows or evolves to other damage types during cycling. The structure properties can then be updated accordingly as a function of cycles, and failure predictions can be generated to determine after how many cycles failure will occur. This is deceptively simple and very hard to implement in a situation where cyclic loads are applied. In fact, determining the onset of damage and monitoring its progress during static loading are still challenging tasks and require advanced simulation methods. It comes to no surprise then that doing that under cyclic loading would be even more challenging.

As it has been stressed repeatedly in this book, see, for example, Chapter 1, predicting the final failure of composite structures is a problem of scales. This is paramount in situations of fatigue loading. One can select the scales at which the model to be developed would be valid. However, no matter how small these scales might be, for example on the order of a fibre diameter, there are always lower scales that are either neglected or not sufficiently captured/understood by the model, for example, the fibre/matrix interface. A fatigue model, therefore, is expected to capture accurately what happens at the scales for which it has been set up, but will have little information about damage creation and accumulation at lower scales. If damage occurs at the scales of validity of the model, then its evolution should be accurately captured. But damage occurring at lower scales could be evolving and the model will not account for it. Thus, properties will be degrading but, until this degradation manifests itself as a change at the scale already captured by the model, it will not be known. The model will be inaccurate without explicitly accounting for this degradation at lower scales.

In view of this, one can envision two requirements for the analytical model to be developed:

- A model that accurately identifies and tracks damage creation and evolution at a preselected scale and at all higher scales.
- A property degradation model that accounts for processes occurring at lower scales than those accounted for by the damage model.

There are two issues of importance associated with such a model. First, scale-up from the length scales selected to full-scale structures must be possible and well defined within the model. Some approximations may be necessary in this context. Second, by necessity, the degradation model will not capture the physics of the phenomena at these lower scales, because if they could be captured, the model would start at these lower scales. This means that the degradation part of the model will be an approximation at best and will introduce inaccuracies.

It is convenient to select a specific property that is a direct outcome of the damage processes at different scales as the characteristic that must be monitored during fatigue cycling and related to fatigue life. This property can be the stiffness [5, 6] or the strength [7] of the structure or some measure of the energy stored [8]. In this chapter, residual strength is selected as the key model parameter. This is a matter of convenience, because using it one can obtain fatigue life as the number of cycles after which the residual strength equals the applied maximum (or minimum) stress. There is no conclusive reason why residual strength should be preferred over any other property. A notional plot showing the relation of cyclic load to residual strength and cycles to failure is shown in Figure 6.4.

Note that the residual strength curve is a function of the cyclic stress σ. There are, therefore, different residual strength curves for different values of σ.

Given the two requirements defined above, an analytical model must be developed that (i) is able to compute the residual strength as a function of damage present in the structure at length scales for which the model is valid and (ii) provides information on how the residual strength evolves (usually degrades) as damage occurs at scales lower than the ones captured in the basic model. These two aspects are addressed in detail in the following sections starting with the lower length scales first.

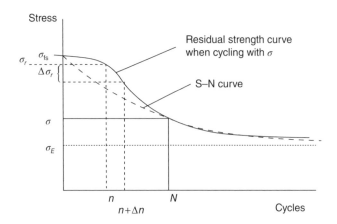

Figure 6.4 S–N and residual strength curves

6.3 Models for the Degradation of the Residual Strength

The residual strength degradation model that addresses processes at scales lower than the ones explicitly accounted for in the analysis will, by necessity, be approximate and phenomenological. It will be based on assumptions that may not apply widely and their verification would require going down to those lower scales the very introduction of which this model is trying to avoid. Recognising these difficulties, it would be useful to have more than one such models for residual strength degradation. These may be complementary, with different models applicable to different situations. They will also show the sensitivity of the overall approach to the use of different models.

6.3.1 Linear Model

In general, for constant amplitude loading at a given stress ratio, the residual strength σ_r will be a function of the static strength σ_{sf} when cycling begins, the applied cyclic load σ, the applied cycles n and the cycles to failure N corresponding to σ. This can be written as

$$\sigma_r = f(\sigma_{sf}, \sigma, n, N) \tag{6.1}$$

One could expand this relation to include the stress ratio as a variable but, for the purposes of the discussion, the stress ratio can be assumed constant and Equation 6.1 is sufficiently general. Applying Equation 6.1 for two different cycle levels n_1 and n_2 gives the corresponding residual strengths:

$$\sigma_{r1} = f(\sigma_{sf}, \sigma, n_1, N) \tag{6.2}$$

$$\sigma_{r2} = f(\sigma_{sf}, \sigma, n_2, N) \tag{6.3}$$

Note that as long as the applied stress σ is constant, and the starting static failure strength σ_{sf} is the same, the cycles to failure N in Equations 6.2 and 6.3 remain the same. Without loss of generality, assume $n_2 > n_1$.

If Equation 6.1 gives the functional form for the residual strength dependence, one could use it to predict σ_{r2} starting from σ_{r1}:

$$\sigma_{r2} = f(\sigma_{r1}, \sigma, n_2 - n_1, N - n_1) \tag{6.4}$$

Equation 6.4 allows starting the degradation process after n_1 cycles. This means that there are $n_2 - n_1$ cycles left to reach n_2 and $N - n_1$ cycles to reach cycles to failure N with σ as the applied cyclic stress. Equation 6.4 is applicable for any pair of values of $n_2, n_1 < N$.

Consider now a special subset of the possible functions f in the form:

$$f(\sigma_{sf}, \sigma, n, N) = \sigma_{sf} + g(\sigma, n, N) \tag{6.5}$$

Using this to substitute in Equations 6.3 and 6.4 gives

$$\sigma_{r2} = \sigma_{sf} + g(\sigma, n_2, N) \tag{6.3a}$$

$$\sigma_{r2} = \sigma_{r1} + g(\sigma, n_2 - n_1, N - n_1) \tag{6.4a}$$

Using now Equation 6.5 to substitute in Equation 6.2 gives

$$\sigma_{r1} = \sigma_{sf} + g(\sigma, n_1, N) \tag{6.2a}$$

This last relation can now be substituted in Equation 6.4a to give

$$\sigma_{r2} = \sigma_{sf} + g(\sigma, n_1, N) + g(\sigma, n_2 - n_1, N - n_1) \tag{6.6}$$

Equating the right-hand sides of Equations 6.3a and 6.6 gives

$$\sigma_{sf} + g(\sigma, n_2, N) = \sigma_{sf} + g(\sigma, n_1, N) + g(\sigma, n_2 - n_1, N - n_1)$$

from which

$$g(\sigma, n_2, N) = g(\sigma, n_1, N) + g(\sigma, n_2 - n_1, N - n_1) \Rightarrow$$

$$g(\sigma, n_2, N) - g(\sigma, n_1, N) = g(\sigma, n_2 - n_1, N - n_1) \tag{6.7}$$

Equation 6.7 gives a lot of insight on the form of the function g. Because it is valid for any values of $n_1, n_2 < N$, if one fixes σ and N, Equation 6.7 would imply that

$$g^*(n_2) - g^*(n_1) = g^*(n_2 - n_1) \tag{6.8}$$

with

$$g^* = g]_{\text{constant } n, N}$$

Equation 6.8 is Cauchy's functional equation, which, if n_1, n_2 are real numbers, can be shown to have only one solution, which has the form:

$$g^*(n) = Cn \tag{6.9}$$

with C an arbitrary constant, provided g^* is continuous at one point at least, or monotonic over an interval or bounded over an interval. Interestingly, the residual strength is not necessarily a monotonic function of cycles. For example, returning to the discussion of Figure 6.3, the residual strength may increase during cyclic loading that creates the longitudinal splits and will decrease after the splits are fully effective. One may also argue that the residual strength may not be a continuous function as damage creation may cause jumps (sudden decreases) to the residual strength. However, it is unlikely that it will be discontinuous everywhere and it takes only one point at which the function must be continuous for Equation 6.9 to hold. In addition, from physical reasoning, the residual strength cannot be unbounded, thus Equation 6.9 is the only possibility.

Combining Equations 6.9, 6.5, 6.2a and 6.1 gives the following expression for the residual strength:

$$\sigma_r = \sigma_{sf} + Cn \tag{6.10}$$

where σ_{sf} is the 'static' strength at the start of the n cycles to be applied. That is, Equation 6.10 is valid at any point in the life of the structure as long as the starting residual strength σ_{sf} is known. Obviously, if the structure has not undergone any cyclic loading, σ_{sf} is the static strength of the pristine structure. The form of Equation 6.10 is such that, before the first cycle, when $n = 0$, the static strength σ_{sf} is reproduced.

The constant C is determined by requiring that after $N-1$ cycles, the residual strength be equal to the maximum (or minimum) applied stress σ. This means that the structure would fail during cycle N, which is, by definition, the number of cycles to failure. Therefore,

$$\sigma_r(N-1) = \sigma \Rightarrow \sigma = \sigma_{fs} + C(N-1) \Rightarrow$$

$$C = \frac{\sigma - \sigma_{fs}}{N-1} \tag{6.11}$$

and, substituting in Equation 6.10 and rearranging, it becomes

$$\sigma_r = \sigma_{sf} - (\sigma_{sf} - \sigma)\frac{n}{N-1} \tag{6.12}$$

For convenience, Equation 6.12 can also be written by normalising stresses with σ_{sf} so that the final answer is obtained as a fraction smaller than one. Thus

$$\frac{\sigma_r}{\sigma_{sf}} = 1 - \left(1 - \frac{\sigma}{\sigma_{sf}}\right)\frac{n}{N-1} \tag{6.13}$$

Equation 6.12 or 6.13 gives the residual strength for a composite structure after n cycles with cyclic stress σ when $\sigma < \sigma_{sf}$ and N is the number of cycles to failure when σ is applied. This functional form was originally postulated by Broutman and Sahu [18]. Here it is given a context and a framework within which it can be shown to follow naturally from the assumptions. It is important to keep in mind that this equation is meant to model damage accumulation and growth at low-length scales for which no information or reliable model is available. This is an important distinction because, as soon as damage manifests itself at the length scales over which analytical models can be developed, different equations would take over. This will be explained in more detail in subsequent sections. In addition, the model described by Equations 6.12 and 6.13 is based on the requirement that the residual strength at any point during the life of the structure must be derived from the residual strength at any previous point in the life using the same mechanism (or functional form as was used here). By definition, because the mechanisms that these equations represent are unknown, there is no deep physical meaning except the boundary conditions that the known residual strength during the first cycle and during cycle N be recovered (σ_{sf} and σ, respectively).

6.3.2 Nonlinear Model

The model presented in the previous section was shown to be linear provided Equation 6.6 is valid. There is no special reason why Equation 6.6 should be valid and, therefore, it would be useful to have an alternate model that does not lead to a linear dependence of the residual strength on cycles. This is based on the approach in Ref. [9].

It is assumed that at any point in the life of the structure, after n cycles, increasing the cycles by an amount Δn will change the residual strength by an amount $\Delta \sigma_r$, which is linearly related to the change in cycles:

$$\Delta \sigma_r = (A\sigma_r + B)\Delta n \tag{6.14}$$

with A and B unknown constants. This is shown in Figure 6.4.

In the limit when $\Delta n \to 0$, Equation 6.14 takes the form:

$$\frac{d\sigma_r}{dn} - A\sigma_r = B \tag{6.15}$$

The solution to Equation 6.15 is

$$\sigma_r = Ke^{An} - \frac{B}{A} \tag{6.16}$$

with K another unknown constant.

The unknown constants in Equation 6.16 are determined by imposing the following three conditions:

1. When $n = 0$, the residual strength equals the static strength of the structure:

$$\sigma_r(n = 0) = \sigma_{sf} \Rightarrow K - \frac{B}{A} = \sigma_{sf} \tag{6.17}$$

2. When $n = N - 1$, the residual strength equals the applied maximum (or minimum) stress σ:

$$\sigma = Ke^{A(N-1)} - \frac{B}{A} \tag{6.18}$$

3. For very large n, the residual strength tends to the endurance limit σ_E (see Figure 6.4). This is the stress below which any cycling will result in infinite life of the structure:

$$\sigma_E = -\frac{B}{A} \tag{6.19}$$

When the assumption that $A < 0$ is made, the exponential in Equation 6.16 vanishes for large n.

Equations 6.17–6.19 form a system of three equations in the three unknowns K, B and A. Solving gives:

$$A = \frac{1}{N-1}\ln\left(\frac{\sigma - \sigma_E}{\sigma_{sf} - \sigma_E}\right) \tag{6.20}$$

$$B = -\frac{\sigma_E}{N-1}\ln\left(\frac{\sigma - \sigma_E}{\sigma_{sf} - \sigma_E}\right) \tag{6.21}$$

$$K = \sigma_{sf} - \sigma_E \tag{6.22}$$

Examination of Equation 6.20 shows that because $\sigma < \sigma_{fs}$ the argument of the natural logarithm in Equation 6.20 is always less than 1. This means that A will be negative, consistent with the assumption following Equation 6.19.

Equations 6.20–6.22 can be placed in Equation 6.16, which, after rearrangement, becomes

$$\sigma_r = (\sigma_{sf} - \sigma_E)\left(\frac{\sigma - \sigma_E}{\sigma_{sf} - \sigma_E}\right)^{\frac{n}{N-1}} + \sigma_E \tag{6.23}$$

In what follows, the endurance limit σ_E will be set equal to zero. This is done for two reasons: First, the cycles of interest here are relatively low and the difference between zero endurance limit and non-zero endurance limit is small. Second, there is evidence that many composite structures do not have an endurance limit [10]. Therefore, with no endurance limit, the residual strength expression becomes

$$\sigma_r = \sigma^{\frac{n}{N-1}}\sigma_{sf}^{\frac{N-n-1}{N-1}} \tag{6.24}$$

or, as a ratio to the starting strength,

$$\frac{\sigma_r}{\sigma_{sf}} = \left(\frac{\sigma}{\sigma_{sf}}\right)^{\frac{n}{N-1}} \tag{6.25}$$

Equations 6.24 and 6.25 are analogous to Equations 6.12 and 6.13. In this model, however, the dependence on cycles n is no longer linear.

As in the case of the linear model in the previous section, there is no clear connection between this model and the physical phenomena taking place at such low-length scales simply because there is no way to obtain such information. The only characteristics of the model that relate to reality are that the rate of change of the residual strength is a linear function of cycles and that the 'boundary conditions' of the residual strength being equal to the starting strength at the beginning of life and the applied stress at the end are recovered.

It is worthwhile, at this point, to compare the two models, the one from this section and the one from the previous section to see how they differ. This comparison is shown in Figures 6.5 and 6.6. Equations 6.13 and 6.25 are used. In both figures, the corresponding curves terminate when σ_r reaches σ, which is at $\sigma_r/\sigma_{fs} = 0.5$ in Figure 6.5 and $\sigma_r/\sigma_{fs} = 0.2$ in Figure 6.6.

Because the cycles to failure, N, are, at this point, unknown, different sets of curves are shown in Figures 6.5 and 6.6 for different values of N: 1 million, 10,000, 1000 and 10. For each value of N, two curves are shown, a continuous line, which is the linear model using Equation 6.13, and a dashed line, which is the nonlinear model, Equation 6.25.

If the applied cyclic stress σ is greater than or equal to 50% of the starting strength σ_{fs}, as in Figure 6.5, the difference between the two models is very small. The continuous and dashed curves for each value of N are very close to each other. In fact, as the ratio σ/σ_{fs} increases from 0.5, the curves come even closer and above 0.65 they are

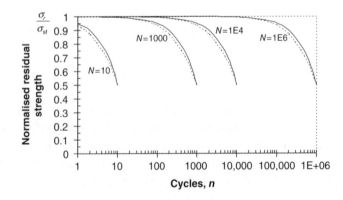

Figure 6.5 Residual strength predicted by the linear and nonlinear models ($\sigma/\sigma_{sf} = 0.5$)

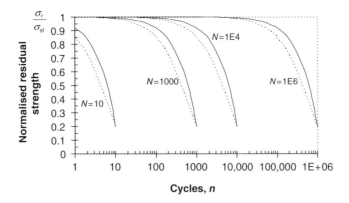

Figure 6.6 Residual strength predicted by the linear and nonlinear models ($\sigma/\sigma_{sf} = 0.2$)

virtually indistinguishable. This means that, for $\sigma/\sigma_{fs} \geq 0.5$, any of the two models can be used without appreciable differences.

If $\sigma/\sigma_{fs} < 0.5$, as in Figure 6.6 where the value 0.2 was used, the differences become more significant and one would have to make a decision which of the two models should be used. It should be noted that the linear model of the previous section is always above the nonlinear model in the present section. Thus, if a behaviour closer to a 'sudden death' behaviour is needed where the residual strength is relatively constant until the cycles to failure are approached and the residual strength drops suddenly, the linear model should be used.

6.4 Model for the Cycles to Failure

The preceding discussion was confined to the residual strength degradation due to damage accumulation and growth at low-length scales, below the ones of interest. During this discussion, N, the cycles to failure, was assumed known. In this section, a model to predict cycles to failure is presented. Again, this model will represent phenomena occurring at low-length scales not represented in the main model that will be presented in the next section.

For simplicity, the discussion will be confined to situations with $R = 0$. The generalisation to other R values will be provided later. If the maximum cyclic stress σ is sufficiently high, then, out of a given population of specimens, those with static strength less than σ will fail during the first cycle. This brings up a useful concept that of the probability that, during any cycle, the applied stress will exceed the strength of a given specimen. This is shown in Figure 6.7.

The probability p that the applied stress σ is greater than the strength of any specimen being tested is given by the ratio of the shaded area in Figure 6.7 divided by the total area under the curve. Figure 6.7 represents the probability density distribution of the residual strength at any time during the fatigue life and the corresponding

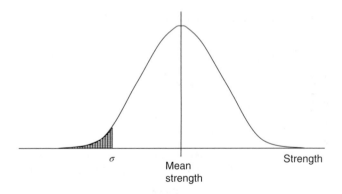

Figure 6.7   Strength probability density distribution

p value will depend on the current status of the structure and the current shape of the strength distribution after n cycles. Thus, p in general depends on n the applied number of cycles.

Obviously, the value of p will depend on the type of statistical distribution of the residual strength. Typically, the static strength of composites can be represented by normal, two-parameter Weibull or lognormal distributions. The type of distribution is not decided arbitrarily. Special purpose tests such as the Anderson–Darling tests [11] are applied to a given set of data to determine the type of distribution and, in several cases, more than one distributions may represent the strength data. However, the one with the highest observed significance level is the most representative and should be used. Usually, the majority of strength data, irrespective of the property, are found to follow a normal distribution. A smaller portion follows a two-parameter Weibull distribution, and an even smaller portion follows a lognormal distribution. A certain percentage of data is found to not be possible to be represented by any of these three distributions. In what follows, the discussion will be confined to two-parameter Weibull or normal distributions but can be generalised to other distributions.

It is important to keep in mind that if the static strength population follows a specific type of distribution, after cycling, the residual strength of the surviving population of specimens may not follow the same type of distribution. Special cases will be described later. It should be obvious by now that the model of residual strength used will have a strong influence on the type of distribution and thus the value of p at any given time during testing. Several cases are distinguished depending on which of the two models presented in the previous section is used for the residual strength as a function of cycles and what statistical distribution represents the strength before cycling.

Assume first that the residual strength is given by Equation 6.12. Then, if the starting strength σ_{sf} follows a normal distribution with mean X and standard deviation s, the residual strength σ_r also follows a normal distribution with mean X_r and standard deviation s_r given by

$$X_r = X - (X - \sigma)\frac{n}{N - 1} \tag{6.26a}$$

$$s_r = s - s\frac{n}{N - 1} \tag{6.26b}$$

Equations 6.26a and 6.26b are based on the fact that if x is a normally distributed variable with mean X and standard deviation s, $ax + b$ is also normally distributed with mean $aX + b$ and standard deviation as.

Then, the probability p that any specimen has strength lower than σ is given by

$$p(n) = \text{cdf}(\sigma, X_r, s_r) \tag{6.27a}$$

where 'cdf' denotes the cumulative distribution function of a normal distribution with mean X_r and standard deviation s_r, evaluated at σ. It is important to note that $p(n)$ given by Equation 6.27a turns out to be constant as a function of n. This can be shown as follows: from the theory of normal distributions, $p(n)$, given by Equation 6.27a, can also be written as

$$p(n) = \frac{1}{2}\left[1 + \text{erf}\left(\frac{\sigma - X_r}{\sqrt{2}s_r}\right)\right] \tag{6.27b}$$

where $\text{erf}(x)$ is the error function of x, with X_r and s_r given by Equations 6.26a and 6.26b. If N is much larger than 1, these expressions for X_r and s_r can be rewritten as

$$X_r \simeq X - (X - \sigma)\frac{n}{N}$$

$$s_r \simeq s - s\frac{n}{N}$$

where the fact that $N \gg 1$ was used.

Then, the argument of the error function in the expression for $p(n)$ becomes

$$\frac{\sigma - X_r}{\sqrt{2}s_r} = \frac{\sigma - X - (X - \sigma)(n/N)}{\sqrt{2}(s - s(n/N))} = \frac{\sigma(1 - (n/N)) - X(1 - (n/N))}{\sqrt{2}s(1 - (n/N))} = \frac{\sigma - X}{\sqrt{2}s}$$

which is independent of cycles. Therefore, the argument of the error function is always constant, and thus $p(n)$ given by Equation 6.27b is independent of cycles.

Equations 6.26b and 6.26a have an important implication about the scatter of the residual strength. Calculating the coefficient of variation (CV) (standard deviation divided by the mean) gives

$$CV_r = \frac{s_r}{X_r} = \frac{s(1 - (n/(N - 1)))}{X_m(1 - (1 - (\sigma/X_m))(n/(N - 1)))} \simeq \frac{s}{X_m}\frac{(1 - (n/N))}{(1 - (1 - (\sigma/X_m))(n/N))} \tag{6.28}$$

where the last expression is an approximation assuming $N \gg 1$.

The ratio of coefficient of variation CV_r, divided by CV, the starting CV (before cycling) is plotted as a function of n/N in Figure 6.8.

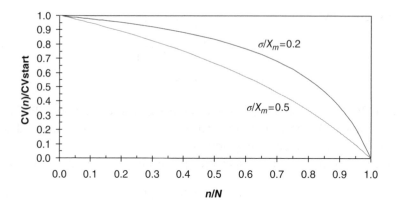

Figure 6.8   Dependence of coefficient of variation of residual strength with cycles

It can be seen from Equation 6.28 and Figure 6.8 that CV_r decreases from its starting value to zero at failure ($n = N$). This means that the scatter of the residual strength decreases according to this model.

If, instead of a normal distribution, the starting strength σ_{sf} follows a two-parameter Weibull distribution with scale parameter β and shape parameter α, the residual strength also follows a two-parameter Weibull distribution with scale parameter β_r and shape parameter α_r given by

$$\beta_r = \beta - (\beta - \sigma)\frac{n}{N-1} \tag{6.29a}$$

$$\alpha_r = \alpha \tag{6.29b}$$

Equations 6.29a and 6.29b are based on the fact that if x follows a two-parameter Weibull distribution with scale parameter β and shape parameter α, then $ax + b$ also follows a Weibull distribution with scale parameter $a\beta + b$ and shape parameter α.

Then, the probability p that any specimen has strength lower than σ is given by

$$p(n) = 1 - e^{-\left(\frac{\sigma}{\beta_r}\right)^{\alpha_r}} \tag{6.30}$$

In this case, it can be shown that as n ranges from 1 to N, $p(n)$ is not constant but increases towards a maximum value of 0.63. This is unlike the previous case where p was constant.

Assume now that the residual strength is given by Equation 6.24. Then, if the starting strength σ_{sf} follows a normal distribution with mean X and standard deviation s, the distribution of the residual strength σ_r is no longer a simple distribution as it involves a normal variate raised to a power:

$$\sigma_{sf}^{\frac{N-n-1}{N-1}}$$

Proceeding with the calculation of p would involve procedures beyond the scope of this chapter. So this option is not pursued further.

If instead of a normal distribution, the starting strength σ_{sf} follows a two-parameter Weibull distribution with scale parameter β and shape parameter α, the residual strength also follows a two-parameter Weibull distribution with scale parameter β_r and shape parameter α_r given by [9]

$$\beta_r = \beta^{\frac{N-n-1}{N-1}} \sigma^{\frac{n}{N-1}} \tag{6.31a}$$

$$\alpha_r = \alpha \frac{N-1}{N-n-1} \tag{6.31b}$$

Equations 6.31a and 6.31b are based on the fact that if x follows a two-parameter Weibull distribution with scale parameter β and shape parameter α then, x^q also follows a two-parameter Weibull distribution with scale parameter β^q and shape parameter α/q.

Then, the probability p that any specimen has strength lower than σ is given by Equation 6.30, which, using Equation 6.31, becomes

$$p(n) = 1 - e^{-\left(\frac{\sigma}{\beta^{\frac{N-n-1}{N-1}} \sigma^{\frac{n}{N-1}}}\right)^{\alpha \frac{N-1}{N-n-1}}} = 1 - e^{-\left(\left(\frac{\sigma}{\beta}\right)^{\frac{N-n-1}{N-1}}\right)^{\alpha \frac{N-1}{N-n-1}}} = 1 - e^{-\left(\frac{\sigma}{\beta}\right)^{\alpha}} \tag{6.32}$$

this is the value of p for the static strength before cycling begins. This means that, for this case, the value of p, irrespective of the value of n, equals its value for the static strength distribution and is, therefore, constant.

Equation 6.31b has similar implications for the scatter of the residual strength as Equation 6.28. The quantity multiplying α on the right-hand side is lower than 1 and increases as n increases. This means that the shape parameter of the residual strength population increases with increasing n. As the shape parameter is a direct measure of the scatter (inversely proportional to the CV), the scatter of the residual strength decreases. This is shown in Figure 6.9. As n increases, the shape parameter increases slowly and, for $n/(N-1) > 0.6$, it starts increasing rapidly towards infinity.

Correspondingly, the scatter follows an analogous decreasing trend.

This trend of decreasing scatter is only specific to the two cases of linear model with a normal distribution for the static strength and the nonlinear model with a two-parameter Weibull distribution for the static strength. It does not necessarily mean that this trend will hold for all composite structures. In fact, whenever the static strength follows a two-parameter Weibull distribution and the residual strength is given by the linear model, Equation 6.29b shows that the shape parameter is constant with cycles and thus the scatter will not change with fatigue loading. As far as experimental evidence is concerned, the data are inconclusive. Yang and Jones [12] have generated test results that showed a reduction in scatter with cycles. On the other hand, results in Ref. [13] show the opposite trend. It is possible that decreasing scatter in residual strength is still compatible with test results even if they showed

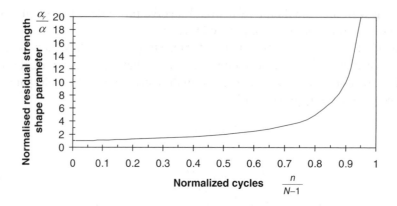

Figure 6.9   Shape parameter of residual strength as a function of cycles

an increase in scatter. The reason is that the models presented here refer to small length scales. As soon as damage manifests itself at longer scales, these models are not valid. Then, it is possible that large-scale damage will increase scatter while small-scale damage may not.

Within this context of small-scale damage, qualitative arguments supporting reduction of scatter in residual strength can be made. As the residual strength decreases with cycles, the specimens with lower strength in the population will fail early. Only stronger specimens survive whose residual strength decreases with cycles. This means the residual strength population will get narrower: at the high end, strength values become smaller, and at the lower end, specimens are removed or, if an endurance stress σ_E is present, strength cannot go below it. This situation is shown schematically in Figure 6.10.

Another reason for reduced scatter is associated with the randomness of strength of composite structures and the fact that structures with flaws tend to exhibit lower scatter. A 'pristine' structure will fail due to some inherent, usually non-detectable, flaw causing a local stress concentration. Depending on where this flaw is relative to areas of high stress gradients within the structure, the failure strength will vary. This results in increased experimental scatter. If, however, there is a flaw created by cyclic loading or exacerbated by cyclic loading, this flaw will drive failure and will overwhelm the effect of other inherent and smaller flaws randomly distributed in the structure. Again in this discussion, the flaw sizes are at very small scale consistent with the assumptions of the models developed in this section.

So far, depending on the model used for the residual strength and the corresponding statistical distribution of the static strength, four possible situations have been analysed as summarised in Table 6.1.

As already suggested, Case 3 will not be pursued further in this chapter because of the complexity of the calculations involved that puts this case beyond the scope of this book. Cases 1 and 4 are consistent with each other in that they predict decreasing

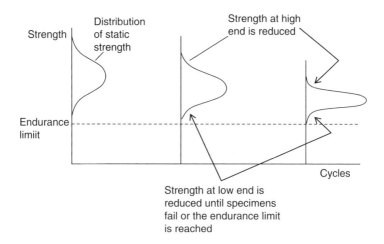

Figure 6.10   Conceptual evolution of strength distribution as a function of cycles

Table 6.1   Residual strength model combinations

Case	Model for residual strength	Statistical distribution of static strength	Scatter during cycling	p: Probability that strength $< \sigma$
1	Linear (Section 6.3.1)	Normal	Decreases	Constant
2	Linear (Section 6.3.1)	Two-parameter Weibull	Constant	Increasing
3	Nonlinear (Section 6.3.2)	Normal	Not examined	Not examined
4	Nonlinear (Section 6.3.2)	Two-parameter Weibull	Decreases	Constant

scatter and constant p with increasing cycles and are the two that will be used from now on. This does not mean that Cases 2 and 3 should not be examined further. In its simplest form, the model proposed here follows directly from Cases 1 and 4.

The main characteristic of interest in the development of a model for cycles to failure is that Cases 1 and 4 have constant p. Physically this means that the damage mechanism occurring at very low-length scales does not change with cycles. This may not be as limiting as it appears. There are examples that suggest that, over a significant number of cycles, and as long as no damage is created at larger length scales, this constant value of p may be a good approximation.

Consider the case of a cross-ply laminate [0/90]s as shown in Figure 6.11. This was first introduced in Section 3.5 and Section 5.7.2.2 within the context of how damage, matrix cracks in particular, may affect overall composite properties. Under cyclic loads, matrix cracks appear in the 90° plies extending across the full thickness of these plies.

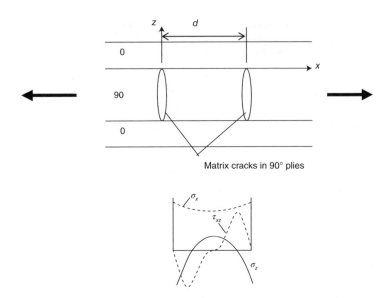

Figure 6.11   Matrix cracks in the 90° ply of a cross-ply laminate under tension–tension fatigue

If one considers as the starting point a situation where the crack spacing is large, then cycling beyond the point shown in Figure 6.11 will show no evidence of damage for a while until a new matrix crack appears between the two shown in Figure 6.11. Between the starting point where cracks are spaced a distance d apart and the appearance of a new crack, damage accumulation is occurring at scales lower than the one at which a matrix crack manifests itself, which is on the order of a few fibre diameters. During this time interval, the axial stress in the 90° ply is transferred through interlaminar shear into the 0 ply. The local stresses in the vicinity of the matrix cracks are shown qualitatively at the bottom of Figure 6.11 (and discussed further in Section 6.6). The residual strength of the laminate as a whole is determined by the strength of the 0 plies in the presence of the interlaminar shear and normal stresses and the locally increased axial stress in the 0 ply. This situation does not change substantially until the next matrix crack appears. This means that the properties do not change significantly. Therefore, we have a situation where the residual strength is relatively constant and the scatter is relatively constant or changes slowly (see region of low n/N in Figures 6.8 and 6.9). It is reasonable to expect that since the residual strength distribution does not change appreciably, p will be relatively constant.

Another example can be drawn from the phenomenon of edge delamination in composite coupons tested under tension. It has been shown in such coupons by O'Brien [14] that the energy release rate for an edge delamination is independent of delamination size. This suggests that, after a delamination starts, further growth of the delamination under cyclic loading does not change the energy release rate as long as the delamination size does not violate the assumptions in Ref. [14]. This growth is not continuous. A number of cycles must be applied before the delamination grows to a new length. During these cycles, again, damage accumulation occurs at scales lower than the delamination size implies. And, during this time, the residual strength stays constant as the strain to failure is independent of the delamination size. As long as the scatter of the residual strength under tension stays relatively constant, p will be relatively constant.

The two examples just presented are not meant to capture completely the physics of the respective situations. They are meant to serve as indicators of situations where the assumption made here for the behaviour of the structure at very low scales may not be entirely unrelated to reality.

Continuing with the conclusion of Cases 1 and 4 that p is constant as a function of cycles, one can consider the possibility that after a number of cycles the structure may fail. It should be emphasised that p is the fraction in the population of specimens whose strength is lower than the applied stress. This is referred to here as the probability that the applied stress will exceed the residual strength of a specimen. Or, $1 - p$ is the probability that the residual strength of a specimen is greater than the applied cyclic stress. Therefore, p is not associated with a specific specimen in the population because the exact residual strength of each specific specimen is not considered here. Only the probability that a given specimen has strength lower than the applied stress is of interest.

Consider now the event that, after a number of cycles, the applied stress σ has exceeded the residual strength of a specimen once. This would mean that the specimen has failed and it is meaningless to consider the event happening more than once. For the purposes of the mathematical model, however, this is allowed. In a more rigorous approach, one can consider that the applied stress gets arbitrarily close to the residual strength of a specimen without ever being equal to it. This would spare the specimen from failure and cycling could continue.

Let P be the probability that after n cycles σ has exceeded σ_r only once. P can be determined as a function of p as follows: over i cycles P_i is the product of the probability p of the event occurring during anyone of these cycles and the probability $(1 - p)$ of it not occurring during all the remaining $i - 1$ cycles:

$$P_i = \underbrace{p(1 - p)(1 - p) \ \dots \ (1 - p)}_{i-1 \text{ times}} = p(1 - p)^{i-1} \tag{6.33}$$

Over n cycles, the probability P is given as the summation of all the individual P_i with i ranging from 1 to n:

$$P = \sum_{i=1}^{n} P_i = \sum_{i=1}^{n} p(1 - p)^{n-1} \tag{6.34}$$

As now p is constant in this simplified model, the summation in Equation 6.34 can be computed by factoring out $p(1 - p)^{n-1}$:

$$P = np(1 - p)^{n-1} \tag{6.35}$$

A plot of P as a function of n is shown in Figure 6.12.

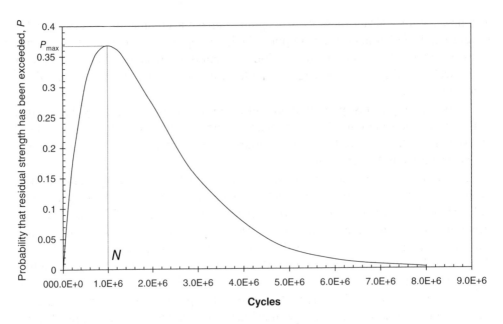

Figure 6.12 Probability of exceeding residual strength as a function of cycles

Figure 6.12 gives a hint on a possible relation between cycles to failure N and P. By looking at Figure 6.12, the most likely moment for failure of the structure is when P is maximised. The value of n that maximises P can be obtained by differentiating Equation 6.35 with respect to n and setting the result to zero:

$$N = -\frac{1}{\ln(1-p)}, \quad R = 0 \tag{6.36}$$

Equation 6.36 gives an estimate of the cycles to failure provided the following assumptions hold:

- No large-scale damage is present; only damage at low scales that cannot be captured with the analytical model chosen.
- The damage present does not change with cycles in a way that would alter the residual strength failure mode.
- Either the residual strength distribution is normal and the linear model of Section 6.3.1 is used, or the residual strength distribution is a two-parameter Weibull and the nonlinear model of Section 6.3.2 is used. In both cases, p is constant with cycles.
- Only one type of loading (tension only or compression only) is present so there is only one value of p during each cycle.

The last assumption must be relaxed so that the approach can be used for cases where $R < 0$. This means that there are two different values of p during each cycle,

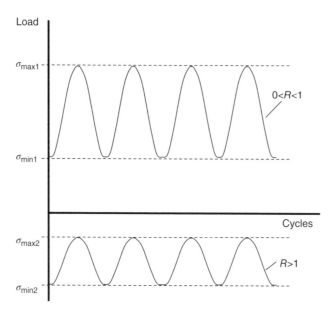

Figure 6.13   Schematic of load cycles for $R > 0$

p_T for the tension portion and p_C for the compression portion. The same procedure as above gives [15]

$$N = -\frac{1}{\ln(1 - p_T) + \ln(1 - p_C)}, \quad R < 0 \tag{6.37}$$

For cases where R is between 0 and 1, tension–tension, or R is strictly greater than 1, compression–compression, further modification is necessary. The reason is that the load excursion does not go to zero for any part of the load cycle. This means that p values based on the residual strength distribution are no longer valid. They must be corrected for the fact that the load excursion is not from 0 to a maximum (or minimum) load but starts from a finite value as shown in Figure 6.13. For example, for $0 < R < 1$, the load excursion is from $\sigma_{\min 1}$ to $\sigma_{\max 1}$ with $0 < \sigma_{\min 1} < \sigma_{\max 1}$. The probability p is not as simple as calculating the probability that $\sigma_{\max 1}$ exceeds the specimen strength minus the probability that $\sigma_{\min 1}$ exceeds the specimen strength. The reason is that the behaviour of a specimen loaded between two finite load values is not equivalent to one specimen loaded from 0 to the maximum value minus the same specimen loaded from 0 to the minimum value.

Avoiding the load portion from 0 to the first load value in the cycle ($\sigma_{\min 1}$ for $0 < R < 1$ and $\sigma_{\max 1}$ for $R > 1$) means that any damage that might occur during that portion of the cycle is avoided. Somehow, the structure must be credited for not under-going this load excursion. This effect is accounted for approximately by adjusting the residual strength distribution. The low end of the distribution, described by the 1% value x_1, is shifted closer to the mean X_m, which is assumed to stay the same.

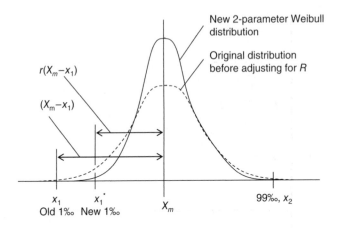

Figure 6.14 Modification of strength probability density function for load cycles not starting at zero

The difference between the mean and the new 1% x_1^* is obtained by scaling the original difference $X_m - x_1$ by a factor r accounting for the portion of the load cycle that is not present:

$$r = 1 - R, \quad 0 < R < 1$$
$$r = 1 - \frac{1}{R}, \quad R > 1 \tag{6.38}$$

This adjustment is shown in Figure 6.14. X_m and the 99th% strength value are kept the same during this adjustment. It is assumed that the resulting skewed distribution is a two-parameter Weibull. Its shape and scale parameters can be obtained as follows:

If the original strength distribution is normal, the 1 and 99% values can be obtained from

$$x_1 = X_m - r(2.326)s \tag{6.39a}$$
$$x_2 = X_m + (2.326)s \tag{6.39b}$$

where the value of 2.326 corresponds to the one-sided tolerance limit factor [16] relating the 99% to the mean and s is the standard deviation.

Then, referring to Figure 6.14,

$$x_1^* = X_m - r(X_m - x_1) \tag{6.40}$$

The shape α_m and scale β_m parameters of the modified distribution are obtained as solutions to the following two equations:

$$\beta_m \left(1 - \frac{1}{\alpha_m} \right)^{\frac{1}{\alpha_m}} = X_m$$

$$e^{-\left(\frac{x_1}{\beta_m} \right)^{\alpha_m}} - e^{-\left(\frac{x_2}{\beta_m} \right)^{\alpha_m}} = 0.98 \tag{6.41}$$

Equations 6.41 are solved iteratively.

If, instead of a normal distribution, the static strength followed a two-parameter Weibull distribution with shape and scale parameters α and β, respectively, the corresponding percentile values are given by

$$x_1 = \beta(-\ln 0.99)^{\frac{1}{\alpha}}$$
$$x_2 = \beta(-\ln 0.01)^{\frac{1}{\alpha}}$$

(6.42)

The corresponding mean value X_m is given by

$$X_m = \beta_m\left(1 - \frac{1}{\alpha}\right)^{\frac{1}{\alpha}}$$

(6.43)

Equations 6.40 and 6.41 are still valid.

With these adjustments, the value of p is updated from that given by Equation 6.27a or 6.30, and Equation 6.36 can be used to obtain the cycles to failure assuming, again, only fatigue processes at scales below the scales of interest are taking place. It should be noted that if $0 < R < 1$ or $R > 1$, as a result of these adjustments, the starting distribution of residual strength is always two-parameter Weibull. Only when $R = 0$, or $R < 1$, where these adjustments are not necessary, it is possible to start with a normal distribution if the residual strength is normally distributed.

For the case where the starting residual strength distribution is two-parameter Weibull and the nonlinear model is used for the evolution of residual strength with cycles, a simple closed form solution can be obtained relating cycles to failure and applied stress. This corresponds to Case 4 in Table 6.1. Equation 6.32 can be used to substitute in the general equation for cycles to failure, Equation 6.36, to obtain after some rearranging:

$$\sigma = \beta\left(\frac{1}{N}\right)^{\left(\frac{1}{\alpha}\right)}$$

(6.44a)

where σ is the maximum (or minimum) cyclic stress, β and α are the scale and shape parameters of the starting residual strength distribution and N is the number of cycles to failure.

As it is, Equation 6.44a does not differentiate between different specimens in the population. It is more like a statement for the entire population and this is why the scale parameter β appears in the right-hand side. This is a direct consequence of the definition of p. To make Equation 6.44a specific to an individual specimen, one must make sure that it recovers the starting residual strength for that specific specimen when $N = 1$. So, for example, for a specimen with mean static strength σ_{sf}, Equation 6.44a becomes

$$\sigma = \sigma_{sf}\left(\frac{1}{N}\right)^{\left(\frac{1}{\alpha}\right)}$$

(6.44b)

Equation 6.44b is easier to use than Equation 6.36 as it is in the more conventional form of S–N curves. It is valid for $R = 0$, that is, tension–tension with minimum stress equal to zero, or $R = \infty$, which corresponds to compression–compression with maximum stress equal to zero. It should be emphasised that it refers to cases where the starting residual strength follows a two-parameter Weibull distribution and the wear-out Equation 6.25 is valid.

A similar expression to Equation 6.44b can be obtained for cases with $R < 0$ by combining Equations 6.32 and 6.37:

$$N = \frac{1}{(\sigma_{max}/\beta_T)^{\alpha_T} + (\sigma_{min}/\beta_C)^{\alpha_C}} \tag{6.45a}$$

where the subscripts 'T' and 'C' refer to tension and compression, respectively, with α and β the respective shape and scale parameters of the starting residual strength distributions and σ_{min} and σ_{max} the respective minimum and maximum stresses of the load cycle.

As in the case of $R = 0$ or ∞, of Equation 6.44a, to make Equation 6.45a specific to an individual specimen, it must be modified to match the starting residual strength for that specimen when $N = 1$. This leads to

$$N = \frac{1}{(\sigma_{max}/\sigma_{sfT})^{\alpha_T} + (\sigma_{min}/\sigma_{sfC})^{\alpha_C}} \tag{6.45b}$$

with σ_{sfT} and σ_{sfC} the mean static strengths in tension and compression, respectively.

This is a bit more complex than Equation 6.44b because it cannot be recast in the conventional S–N form where the stress is isolated on the left-hand side of the equation. It has an important characteristic in that it shows that, in general, both the tensile, σ_{msx}, and compressive, σ_{min}, magnitudes of the cycle affect the cycles to failure N.

6.4.1   Extension to Spectrum Loading

The wear-out model presented in the previous section was developed for constant amplitude loading. It can be extended to spectrum loading following the approach in Ref. [17].

The extension uses the residual strength as a key parameter that allows 'equivalencing' one type of loading with another. Two different specimens that have undergone different type of cyclic loading will be considered equivalent if, at the end of cyclic loading, they have the same residual strength. This does not mean that the specimens have the same type of damage. It is well known that different types of damage may lead to the same residual strength. For example, for any laminate with BVID (barely visible impact damage), there is a hole size that results in the same compression strength as the compression-after-impact (CAI) strength (see also Section 5.3). This equivalence only means that, at a given point during their fatigue lives, the two specimens have the same residual strength for a given type of loading. Therefore, the approach here is not expected to be valid for spectrum loading situations where the loading type changes (for example, compression vs shear) or where, for the same type of loading but different magnitude, substantially different types of damage result in the specimen.

The problem then is reduced to first determining the residual strength of a specimen after it has been cycled at one load level for a certain number of cycles and then determining the number of cycles that would result in the same residual strength after cycling at a different load level. For example, assume a specimen undergoes n_1 cycles at cyclic stress σ_1 and then n_2 cycles at cyclic stress σ_2. After the first set of n_1 cycles, the residual strength can be obtained using, for example, Equation 6.24 as

$$\sigma_{r1} = \sigma_1^{\frac{n_1}{N_1-1}} \sigma_{sf}^{\frac{N_1-n_1-1}{N_1-1}} \tag{6.46}$$

where N_1 are the cycles to failure if cyclic loading at σ_1 were continued indefinitely.

The same residual strength σ_{r1} can be obtained by cyclic loading with cyclic stress σ_2 instead of σ_1. Letting N_{2u} denote the number of cycles at load σ_2 to reach residual strength σ_{r1}, using Equation 6.24 again, gives

$$\sigma_{r1} = \sigma_2^{\frac{N_{2u}}{N_2-1}} \sigma_{sf}^{\frac{N_2-N_{2u}-1}{N_2-1}} \tag{6.47}$$

where instead of n_1 and N_1, N_{2u} and N_2 were used, respectively.

Equation 6.47 can be combined with Equation 6.44b to solve for N_{2u}:

$$N_{2u} = (N_2 - 1)\frac{\ln \sigma_{r1} - \ln \sigma_{sf}}{\ln \sigma_2 - \ln \sigma_{sf}} = (N_2 - 1)\frac{\ln(\sigma_{r1}/\sigma_{sf})}{\ln(\sigma_2/\sigma_{sf})} \tag{6.48}$$

This result can be combined in Equations 6.47 and 6.44b to solve for N_{2u} [17]:

$$N_{2u} = \frac{N_2 - 1}{N_1 - 1} n_1 \frac{\ln N_1}{\ln N_2} \tag{6.49}$$

So, if instead of n_1 cycles with load σ_1, N_{2u} cycles with load σ_2 were applied, the residual strength at the end of cyclic loading would be the same. This means that the original requirement of n_1 cycles with load σ_1 followed by n_2 cycles with load σ_2 can be replaced with $N_{2u} + n_2$ cycles all with load σ_2. Therefore, the residual strength at the end of the two segment loading can be obtained by using Equation 6.24 with load σ_2 applied for $N_{2u} + n_2$ cycles. This leads to

$$\sigma_{r2} = \sigma_2^{\frac{n_2}{N_2-1}+\frac{n_1}{N_1-1}\frac{\ln N_1}{\ln N_2}} \sigma_{sf}^{1-\left(\frac{n_2}{N_2-1}+\frac{n_1}{N_1-1}\frac{\ln N_1}{\ln N_2}\right)} \tag{6.50}$$

The procedure can be repeated with the order of load application reversed and it can be shown [17] that the residual strength is the same as that given by Equation 6.50. So within the context of the present model, changing the order of load application does not change the residual strength. However, as will be shown later, the cycles to failure will, in general, change.

Before proceeding, it is worth emphasising that this model is based on the assumptions that the failure mode does not change as the loads change and that the cycle-by-cycle probability of failure p is constant. It is easy to envision situations where these assumptions are not valid. However, in most cases, this occurs at scales longer than the ones examined here, which will be covered by the discussion in Section 6.6.

One can show by induction [17] that if instead of the two segments discussed here, m load segments were applied, the equivalent constant amplitude cycles N_{mu} at load σ_m right before the mth segment starts and the corresponding residual strength are given by

$$N_{mu} = \frac{(N_m - 1)}{\ln N_m} \left[\sum_{i=1}^{m-1} \frac{n_i}{N_i - 1} \ln N_i \right] \tag{6.51}$$

$$\sigma_{rm} = \sigma_m^{\frac{n_m + N_{mu}}{N_m - 1}} \sigma_{sf}^{\frac{N_m - (N_{mu} + n_m) - 1}{N_m - 1}} \tag{6.52}$$

Equation 6.51 gives the number of cycles under stress σ_m, which is the load at the last segment that would lead to the same residual strength, given by Equation 6.52, if the different load segments were applied. Therefore, if N_m is the number of cycles to failure when constant amplitude σ_m is applied, the remaining life, n_m, before segment m starts is

$$n_m = N_m - N_{mu} \tag{6.53}$$

Using Equation 6.51 to substitute in Equation 6.53 and rearranging leads to

$$\frac{1}{\ln N_m} \left[\frac{n_m}{N_m} \ln N_m + \frac{(N_m - 1)}{\ln N_m} \sum_{i=1}^{m-1} \frac{n_i}{N_i - 1} \ln N_i \right] = 1 \tag{6.54a}$$

or, if $N_i > 20$, in which Case 1 can be neglected compared to N_i, the following expression results:

$$\frac{1}{\ln N_m} \sum_{i=1}^{m} \frac{n_i}{N_i} \ln N_i = 1 \tag{6.54b}$$

Equation 6.54a or its approximation Equation 6.54b is the condition for failure after m load segments and is analogous to the traditional Miner's rule used mostly for metals:

$$\sum_{i=1}^{m} \frac{n_i}{N_i} = 1 \tag{6.55}$$

In Equations 6.54a, 6.54b and 6.55, N_i are the cycles to failure if only the load σ_i of the ith segment were applied.

Unlike Miner's rule which is a linear relationship, Equation 6.54a is nonlinear. This means that changing the order of application of load segments has no effect on the cycles (or segments) to failure if Miner's rule is used but can have a big effect if Equation 6.54a or 6.54b is used.

A second important difference between the present model and Miner's rule is that, in Miner's rule, n_i/N_i always adds to 1 while, in the present model, it can add to a

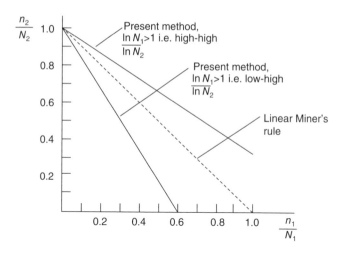

Figure 6.15   Comparison of present method to Miner's rule predictions

quantity greater or lower than 1 depending on the situation. Taking the two-segment case for simplicity, Equation 6.54b leads to

$$\frac{n_2}{N_2} = 1 - \frac{n_1}{N_1}\frac{\ln N_1}{\ln N_2} \tag{6.56}$$

Clearly, the slope of n_1/N_1 in the right-hand side is different than 1, except when $N_1 = N_2$. A plot of Equation 6.56 is shown in Figure 6.15.

If the high load segment is applied first followed by the low load (high–low sequence), the present method predicts that the sum of n_i/N_i is always greater than 1. If the sequence is reversed, low–high sequence, the present method predicts that the sum of n_i/N_i is lower than 1. There is experimental evidence that partially supports these conclusions. Broutman and Sahu [18] report that 6 out of 12 high–low cases had Miner's sum greater than 1, supporting the conclusions of the present model, and 10 out of 12 low–high cases had sum less than 1 again supporting the predictions of the present model.

Another implication of the present model is that the cycles (or blocks) to failure for a given spectrum loading will not change if the last load segment remains the same. That is, if all but the last segment are rearranged in any order, the blocks to failure will be the same. This is a result of the fact that the present model yields the same residual strength right before the last load segment begins irrespective of the order of the segments up to that point. Only if the last segment were changed, would the number of blocks to failure change. At this point, there is no evidence that this conclusion is valid but there is no statistically significant evidence that it is not.

This conclusion can be used to lead to a relatively simple expression for the number of blocks to failure given a particular spectrum. It is recognised that failure will occur when the applied stress during a certain load segment equals the corresponding residual strength of the specimen. Then, the number of load segments to failure M_{fail} when the applied stress equals σ_i with σ_i the maximum (or minimum) stress during segment i can be shown to be [17]

$$M_{\text{fail}} = \frac{1}{K_m} \left(\ln N_i - \frac{n_m}{N_m - 1} \ln N_m \right) \tag{6.57}$$

where N_m and n_m are the cycles to failure and applied number of cycles for the last segment in the spectrum, respectively, N_i is the number of cycles to failure when only σ_i is applied, which can be determined from Equation 6.44b, and K_m is given by

$$K_m = \sum_{i=1}^{m} \frac{n_i}{N_i} \ln N_i \tag{6.58}$$

Equation 6.57 is applied successively for each σ_i in the spectrum and the lowest value of M_{fail} is the sought-for prediction for blocks to failure.

6.5 Residual Strength and Wear-Out Model Predictions Compared to Test Results

The discussion so far concentrated on the development of models for the residual strength and cycles to failure that address phenomena at scales lower than the ones at which explicit modelling of damage creation and evolution will take place. These constitute, therefore, wear-out models that allow estimation of how structural properties change as a function of cycles even if there is no apparent change to the damage present in the structure as assessed by the fatigue model at longer scales. The fatigue model at longer scales will be presented in Section 6.6. Before presenting that model, however, it is useful to evaluate the predictive capability of the present models, within the limitations of the assumptions made and the test results available in open literature.

6.5.1 Residual Strength Predictions Compared to Test Results

As the model developed in this chapter depends heavily on a good prediction of residual strength as a function of cycles, the first comparison with test results is for the residual strength. Two sets of comparisons are made. The first uses the test results from Yang and Jones [12]. In these tests, carbon/epoxy specimens were cycled at a given load level for a number of cycles at $R = 0.1$. At the end of the test, they were loaded statically to failure and the residual strength was recorded. The predictions of models 2 and 4 of Table 6.1 are compared to test results in Table 6.2.

The first two columns in Table 6.2 defined the number of cycles and maximum applied stress. The third column in Table 6.2 gives the predicted cycles to failure using Equation 6.36. Note that because R is nearly 0, the correction of Equation 6.38 was found to have a negligible effect on the residual strength prediction and was

Table 6.2 Residual strength predictions compared to test results from Ref. [12]

Applied cycles, n	Applied stress σ_{max} (MPa)	N_f to failure under σ_{max} (Equation 6.36)	Test residual strength from Ref. [12] (MPa)	Prediction residual strength (Equation 6.24)	% Difference of Equation 6.24 from test	Prediction residual strength (Equation 6.12)	% Difference of Equation 6.12 from test
1,100	298.1	2,663	392.3	363.7	−7.3	368.6	−6.0
12,100	268.3	30,918	379	351.6	−7.2	359.6	−5.1
137,500	238.4	483,964	363.7	356.5	−2.0	367.2	+1.0
150,000	232.9	833,292	348.8	376.4	+7.9	384.9	+10.3
900	290.7	4,781	390.6	390.6	−3.6	394.3	+1.0

not used. The residual strength obtained experimentally in Ref. [12] is given in the fourth column of Table 6.2. The predicted residual strength from the non-linear model, Equation 6.24, and the linear model, Equation 6.12, are given in columns 5 and 7 with the corresponding percent differences from test results in columns 6 and 8. As can be seen from the percent differences, the residual strength predictions are very good and within, at worst, 10% of the test results. The predictions from the linear model are slightly better with the worst case differing from test by 8%.

A second comparison was done on fibreglass with results published by Broutman and Sahu [18]. Here again, a certain applied load σ was selected as a fraction of the static tensile strength and the number of cycles n was applied as a fraction of the experimentally determined mean cycles to failure N. The results are summarised in Table 6.3. The residual strength values for test and predictions are given as percentages of the static strength.

Table 6.3 Normalised residual strength predictions compared to test results from Ref. [18]

n/N	Test [18]	Prediction Eq. 6.25	Prediction Eq. 6.13	n/N	Test [18]	Prediction Eq. 6.25	Prediction Eq. 6.13
	$\sigma/F_{tu} = 0.862$				$\sigma/F_{tu} = 0.754$		
0.20	93.0	97.1	97.2	0.20	91.5	94.5	95.1
0.51	94.5	92.7	92.9	0.55	89.0	85.6	86.5
	$\sigma/F_{tu} = 0.646$				$\sigma/F_{tu} = 0.538$		
0.175	90.5	92.6	93.8	0.15	83.0	91.1	93.1
0.56	83.5	78.3	80.2	0.35	81.0	80.5	83.8

As the ratio n/N was given from the test results, Equation 6.25 was modified for large N as follows:

$$\frac{\sigma_r}{\sigma_{sf}} = \left(\frac{\sigma}{\sigma_{sf}}\right)^{\frac{n}{N-1}} \simeq \left(\frac{\sigma}{\sigma_{sf}}\right)^{\frac{n}{N}}$$

which eliminated the need for predicting N. This was then used to obtain the first set of predictions in Table 6.3. The second set was obtained from Equation 6.13 with an analogous approximation:

$$\frac{\sigma_r}{\sigma_{sf}} = 1 - \left(1 - \frac{\sigma}{\sigma_{sf}}\right)\frac{(n-1)}{N-1} \simeq 1 - \left(1 - \frac{\sigma}{\sigma_{sf}}\right)\frac{n}{N}$$

valid when $N \gg 1$.

The predictions are in good agreement with test results with the worst deviation of +9.7% for the nonlinear model of the residual strength, Equation 6.25 and +12.1% for the linear model, Equation 6.13. The worst predictions occur for $\sigma/F_{tu} = 0.538$ and $n/N = 0.15$.

The results in Tables 6.2 and 6.3 suggest that the residual strength model based on Equations 6.12 or 6.25 gives reliable predictions.

6.5.2 Cycles to Failure Predictions Compared to Test Results (Constant Amplitude)

Comparisons of analytical predictions to test results for various materials, geometries and loadings are presented in this section. The first comparison is with tests done by Broutman and Sahu [18] at $R = 0.05$ on fibreglass cross-ply laminates. The comparison is shown in Figure 6.16. The test data include the 10% and 90% and mean data point. The predictions are obtained with Equation 6.44b.

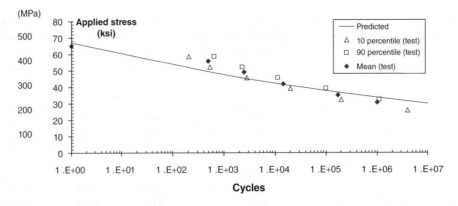

Figure 6.16   Analytical predictions versus test results for fibreglass cross-ply laminates

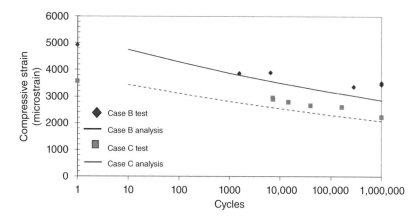

Figure 6.17 Analytical predictions versus test results for sandwich with impact damage

Very good agreement is observed between predictions and test results except for high applied loads where the predictions are conservative. For cross-ply laminates at high cyclic loads, the 90° plies are saturated with matrix cracks. This alters the load transfer through the specimens and is not taken into account in the present model. At lower loads, the matrix crack density in the 90° plies is not as high and the present model accounting for wear-out at small scales gives reasonable predictions. The effect of matrix cracks will be discussed in more detail in Section 6.6.2 when the fatigue model at longer length scales is presented.

A second comparison is done in Figure 6.17 for sandwich with carbon/epoxy fabric facesheets. These are the same as Cases B and C in Section 5.6. There was no damage growth during the tests at least when inspected visually or with tap test. So this is a case where damage accumulation occurred at scales below those of the damage present and, therefore, the model is applicable. The predictions are in reasonable agreement with test results. More test data would be needed for more accurate evaluation. Also, higher resolution inspection during testing would be needed to see whether there were any damage growth that visual and tap inspection did not identify. It is suspected that part of the discrepancy is because some damage growth at longer scales (a few millimetres) may have taken place during the test.

The next comparison is in Figure 6.18 for $[(\pm 45/0_2)_2]s$ T800/5245 bismaleimide for two different R values. Test data are taken from Ref. [19]. Very good agreement is observed between test results and analytical predictions.

While the results in Figures 6.16–6.18 are encouraging, they are by no means sufficient. The model as used is limited because it does not account for damage creation and growth at longer scales than a few fibre diameters. Longer scales will be covered by the second part of the model that will be presented in Section 6.6. A comparison that easily shows the limitations of the present method is shown in Figure 6.19.

This is a double lap-bolted joint with T300/914 material and layup: $[0_2/\pm 45/0_2 /\pm 45/90]s$ base plate with $[0_2/45/9/-45/90]s$ doublers. Loading was tension–

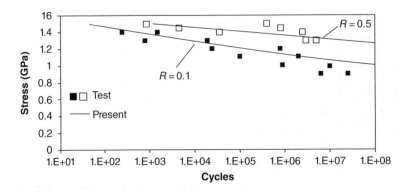

Figure 6.18   Analytical predictions versus test results for bismaleimide laminates

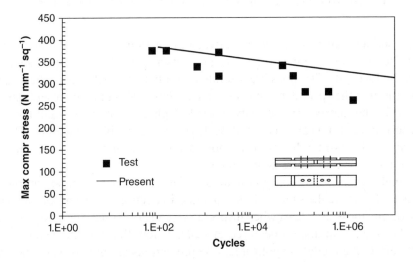

Figure 6.19   Analytical predictions versus test results for a bolted joint with $R = -1.66$

compression with $R = -1.66$. Test data are from Ref. [20]. Here, the discrepancy between analytical predictions and test results is significant. It is to be expected as the bolted joint has complex damage creation and evolution mechanisms, none of which is accounted by the model as presented so far.

More comparisons for different materials and loadings giving a better feel for situations where the model is not sufficiently accurate can be found in Ref. [15].

6.5.3   Cycles to Failure Predictions Compared to Test Results (Spectrum Loading)

Two sets of comparisons are provided in this section for spectrum loading situations. The first is shown in Figure 6.20 and the test data are taken from Ref. [18]. The analysis method from Section 6.4.1 was used. More details about implementing this analysis

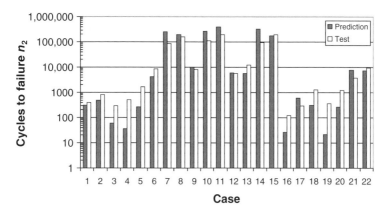

Figure 6.20   Analytical predictions compared to test results under two-segment spectrum loading

can be found in Ref. [17]. Each case in Figure 6.20 corresponds to a different scenario where specimens were first loaded for a number of cycles at one stress level and then loaded under a different stress level to failure. The y-axis in Figure 6.20 shows the cycles of the second stress level that are required to cause failure.

In general, there is good agreement between analytical predictions and test results except for cases 3, 4, 5, 18, 19 and 20. In all these cases, the applied load was high, either 86% or 75% of the mean static strength. For these high loads, Figure 6.16, which corresponds to exactly the same material, layup and loading, showed that the predicted S–N curve departs from the test results. So this discrepancy under spectrum loading is to be expected. It should be noted also that the y-axis scale in Figure 6.20 is logarithmic (to fit all cases in one plot), which means the discrepancies are more significant than they appear.

The second comparison is shown in Figure 6.21 with test data taken from Ref. [21]. This is a situation where a $[(\pm 45/0_2)_2]$s T800/5245 laminate was tested under spectrum loading. The spectrum consisted of four segments each with $R = 0.1$ and with maximum stress corresponding to 78%, 72%, 66% and 60% of the static strength. These were mixed in different orders for each of the six cases shown in Figure 6.21 and repeated until failure.

It appears from Figure 6.21 that the analytical predictions are in good agreement with tests because they are always within the experimental scatter that is quite large. However, this is a bit misleading as other models could give equally good if not better predictions. In addition, the predictions of the present model consist of only two different values for all six cases because there were only two different last segments in the spectrum across all six cases. As was mentioned in Section 6.4.1, if the last segment remains the same, the present model gives the same prediction even if the order of the previous segments changes.

Finally, it should be pointed out that the model for spectrum loading is quite limited as is. It does not account for larger scale damage that is very important

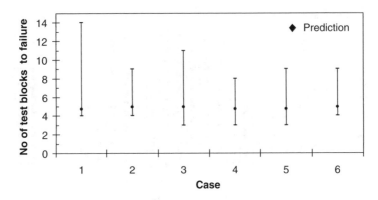

Figure 6.21   Analytical predictions versus test results for four-segment spectrum loading

in situations where the tension-dominated cycles are followed or preceded by compression-dominated cycles. Ways to incorporate these effects on a cycle-by-cycle basis will be presented in the next section.

6.6   A Proposal for the Complete Model: Accounting for Larger Scale Damage

Section 6.4 presented a model for assessing fatigue lives if the damage processes are at very low-length scales, below the length scale at which the damage model is created. For example, if the damage of interest consists of matrix cracks, delaminations and broken fibres, the scale of interest is that of a few fibre diameters. Then, the model presented in Section 6.4 becomes a wear-out model that tracks damage creation and accumulation at scales near or lower than one fibre diameter.

The damage model will depend on the laminate layup, loading and part geometry. A laminate with a hole, for example, would require a different damage model than a laminate with a ply drop developing a delamination at the location where the ply was dropped. Therefore, the specifics of the damage model will change from one case to the next. Here, a generic approach will be presented and, in subsequent sections, examples will be given to show how it can be applied to different cases.

It is assumed that the state of the structure at the beginning of cyclic loading is accurately known. This could be a nominally 'pristine' structure or a structure that has already seen some cyclic loading. The (residual) strength of the structure before cyclic loading begins must be known and it will be a function of whatever damage state is present. This makes it necessary to have accurate static analysis models available. In addition to the static strength, the scatter around that static strength and its associated statistical distribution must be known.

Consider such a structure to which cyclic loading is applied with a given R value. Conceptually, the process is relatively straightforward and is based on tracking the residual strength as a function of cycles. To illustrate the approach, consider a loading situation with $R < 1$ (tension–compression).

6.6.1   First Cycle, Tension Portion

Apply the tension part of the cycle. The load increases from 0 to σ_{max}. If this load causes any damage, the stiffness and strength properties of the structure are updated to reflect the new damage and a new residual strength is computed with its associated statistical distribution. If this load causes no damage in the structure, it is reasonable to assume that damage is occurring at scales lower than the model can capture and the wear-out model of Section 6.4 is used to update the residual strength of the structure. The 'updated structure' is the new structure with which the compression part of the cycle will begin.

It is important to note here that this process must, in general, track more than one residual strength values depending on the loading. For example, for the tension–compression loading situation, both the residual strength under tension and the residual strength under compression must be determined at the end of the tensile portion of the load cycle.

A simple example would be the case of a unidirectional laminate with the fibres aligned with the direction of the loading. During the tension portion of the cycle, some fibres, the weaker fibres, will fail. Fibre failure here refers to complete failure of the fibre and not the presence of fibre cracks around which the matrix can carry load, so broken fibres still transfer load. Essentially, 'broken fibres' in this context are fibres with multiple breaks along their length, and, which fibres, can no longer carry load. Then, the load on these failed fibres is redistributed to the remaining fibres. As a result of this load increase on the intact fibres, some more fibres may break. Load is further redistributed until no more fibres are left to take the load, resulting in complete failure, or, there are some fibres left able to carry the load. The fibre volume is now adjusted to account for the fact that some fibres have failed. As a result, the stiffness and strength for the resulting structure are updated.

6.6.2   First Cycle, Compression Portion

The damage state, stiffness and strength of the structure at the end of the tension portion of the cycle above form the starting state of the structure for the compression portion. In particular, the residual strength in compression is the key variable.

Apply the compression part of the cycle. The load is decreased from 0 to σ_{min}. If this load causes any damage to the structure, the stiffness and strength properties are adjusted to reflect the new damage created. New residual strength values for tension and compression are calculated along with their associated statistical distributions. If this part of the load cycle causes no damage, again, it is postulated that damage is occurring at lower scales than the ones captured in the damage model and the wear-out model of Section 6.4 is used to update stiffness and strength properties. The updated state of the structure will be used to start the next tension portion for the next cycle.

As with the tension cycle above, the new residual strength not only for compression but also for tension must be determined, reflecting the state of the structure at the end of the compression part of the cycle. This can be used for the tension portion of the next cycle.

Continuing with the unidirectional laminate example, at the end of the compression part of the cycle, the portion of the fibres that failed is determined and their load is redistributed to the remaining fibres. As some more fibres may fail, this process continues until either all fibres have failed resulting in final collapse, or some fibres remain intact, able to carry the load. In the latter case, the new fibre volume is calculated and the new stiffness and strength properties determined. The structure is ready to undergo the next load cycle.

6.6.3 Subsequent Load Cycles

The procedure described above for the first cycle is repeated for each subsequent cycle. At the beginning of each new cycle, the structure has been updated in accordance with the damage state or wear-out model predictions. The process terminates when either all plies have failed, or one of the residual strengths, tensile or compressive, equals the corresponding applied cyclic stress.

6.6.4 Discussion

The conceptual model just described attempts to combine (i) processes at low-length scales that affect macroscopic properties such as strength and stiffness and (ii) damage creation and evolution at macroscopic length scales. In this context, the wear-out model of Section 6.4 can no longer be used to predict cycles to failure. It only serves to predict changes in residual strength resulting from processes in microscopic scales. In a sense, it allows estimation of strength reductions from one cycle to the next due to material wear-out. Of course, the wear-out model makes use of cycles to failure N, see, for example, Equations 6.36 and 6.44b but this N refers to cycles to failure as a result of wear-out and not damage creation and growth at longer scales. Needless to say that N, as well as p, now changes with every cycle. But if there is no larger scale damage, p and N are assumed constant until such damage occurs and p and N are updated accordingly.

For the approach to be used effectively, an analytical model that accounts for damage created and its effect on residual strength is necessary. This means that good static models are needed that can predict different types of damage and their interaction and growth such as matrix cracks, delaminations and fibre breakage. Currently, such models are computationally very intensive [22–25] and are difficult to use on a cycle-by-cycle evaluation of the structure. For the time being, less accurate approximations of damage onset combined with progressive failure analysis such as first ply and post-first-ply failure criteria can be used instead [26, 27].

One of the challenges of the proposed model, of any fatigue model for that matter, even equipped with some of the best numerical simulation models as in Refs [22–25], is damage growth under fatigue loading. As mentioned in Section 6.1, damage in composites takes many different forms and growth may refer to growth of any and all of these different forms. Self-similar damage growth is rare.

As more accurate and less time-consuming progressive damage analysis models become available, they can be used in the present model to dynamically update the residual strength not at the end of the tensile and/or compressive portion of a load cycle but during the cycle itself. This will allow more accurate tracking of damage growth.

One final comment about the use of the wear-out model in this context is in order. The model of Section 6.4 assumes that, barring any longer scale damage events, the residual strength decreases with cycles. This means that, in its current form, it will not capture situations where residual strength increases (see also earlier discussion in relation to Figure 6.3). This is to be expected because, for the case of a hole in a unidirectional specimen under tension, there is larger scale damage, namely the matrix splits at the edges of the hole that occur, whose effect must be accounted for by the longer scale model.

The main advantage, but also challenge of the present model is that it does not need any fatigue testing to determine baseline S–N curves or to curve-fit any model parameters. As such, it can be quite powerful. In its simplest form, described in the next two sections, it can be used as a preliminary design tool to help identify fatigue-resistant configurations and minimise fatigue testing. In a more complex form, when updated with some of the more powerful simulation models, it will be even more useful in a design and analysis environment.

6.6.5   Application: Tension–Compression Fatigue of Unidirectional Composites

The model described in the previous section is now applied to unidirectional composite coupons under tension—compression with $R = -1$.

6.6.5.1   Predictions

The following assumptions are made:

- All fibres in a coupon have the same stiffness.
- If some fibres fail, all intact fibres are equally loaded (no shear lag through matrix).
- The strength of the fibres in a coupon follows a normal distribution with known mean and standard deviation.

The last two assumptions are of some importance. The third assumption recognises that not all the fibres in a coupon have the same strength. This means that, during cyclic loading, the weaker fibres remaining in the coupon will fail and load will be redistributed to the stronger fibres. As implied by the second assumption, the load will be equally distributed to the surviving fibres. This is conservative as it has been shown by Qian [28] that, for hexagonally or near hexagonally closed packed fibre patterns, the stress concentration factor in fibres immediately adjacent to a broken fibre is less than 1.16.

At the beginning of the cyclic loading, the following quantities are assumed known:

- Mean fibre tension strength X_f^t and corresponding standard deviation s_f^t.
- Mean fibre compression strength X_f^c and corresponding standard deviation s_f^c.
- Fibre volume v_f.

For failure under tension, a detailed model would account for any fibre waviness, variations in fibre diameter, inconsistencies at the fibre/matrix interface and so on. A model accounting for fibre waviness and fibre diameter variation was introduced in Ref. [29]. For compression, a model accounting for fibre kinking and the effect of fibre waviness in compression strength [30] should be used. Here, a simpler rule of mixtures will be used in order to obtain preliminary predictions and demonstrate the approach.

Assume a simple rule of mixtures relating fibre strength to composite strength:

$$\sigma_{\text{fail}} = X_f v_f \tag{6.59}$$

where X_f is the fibre strength and v_f is the (current) fibre volume. The values σ_{fail} and X_f change depending on whether the fibre and coupon are under tension or compression. It is assumed that the matrix has a negligible contribution to the composite strength. This is why Equation 6.59 contains no matrix contribution.

First Cycle Tension Portion
Apply the tension part of the cycle. The applied stress on the fibres is given by a relation analogous to Equation 6.59:

$$(\sigma_{\text{fa}}^t)_1 = \frac{\sigma}{v_f} \tag{6.60}$$

All fibres with strength less than $(\sigma_{\text{fa}}^t)_1$ fail. The fraction of fibres that failed f_1^t is given by

$$f_{1\text{new}} = \text{cdf}((\sigma_{\text{fa}}^t)_1, X_f^t, s_f^t) \tag{6.61}$$

where 'cdf' denotes the cumulative distribution function with mean X_f^t and standard deviation s_f^t.

The remaining fibres are now all (equally) loaded by the new stress:

$$(\sigma_{\text{ft}})_{\text{new}} = \frac{(\sigma_{\text{fa}}^t)_1}{1 - f_{1\text{new}}} \tag{6.62}$$

Return now to Equations 6.60 and 6.61 and check whether more fibres fail due to the load redistribution. Repeat this process until either convergence of $f_{1\text{new}}$ to $f_1' < 1$ or all fibres fail.

Based on the converged value of f_1, calculate the new fibre volume of the specimen:

$$v_{f1}^t = (1 - f_1^t)v_f \tag{6.63}$$

Calculate new fibre strengths. The individual fibres that have not failed so far do not have any damage that can be assessed by the analysis model in Equations 6.60–6.63. So their strength at the end of the first cycle will be updated using the wear-out model: First, calculate the number of cycles N to fibre failure using Equation 6.36, rewritten for the tension portion as

$$N = -\frac{1}{\ln(1 - p_{T1})} \tag{6.64}$$

with p_{T1} the probability that the applied stress σ is greater than the tensile strength of a specimen, given by

$$p_{T1} = \text{cdf}\left(\frac{\sigma}{v_{f0}}, X_f^t, s_f^t\right) \tag{6.65}$$

It is important to note that N, given by Equation 6.64, is not the cycles to failure under subsequent loading. It would be the cycles to failure only if under subsequent loading no more fibres failed and no other (detectable) damage appeared in the specimen.

The residual strength of the fibres is updated using the wear-out model. To calculate the residual strength after one cycle, one can use Equation 6.12 or 6.24. Here, the linear model of Equation 6.12 is used for illustration purposes. After one cycle, the residual strength follows a normal distribution with mean and standard deviation given by Equations 6.26a and 6.26b, respectively:

$$\text{mean}_1 = \left(\frac{N-2}{N-1}\right)(\text{mean})_0 + \frac{\sigma}{N-1}$$

$$\text{stdev}_1 = \left(\frac{N-2}{N-1}\right)(\text{stdev})_0 \tag{6.66}$$

where '0' refers to the situation before the first cycle.

Applying Equation 6.66 to fibre strength, the mean and the standard deviation of the tension strength of the fibres are given by

$$X_{f1}^t = \left(\frac{N-2}{N-1}\right)X_f^t + \frac{(\sigma_{fa}^t)_1}{(1 - f_{1\text{new}})(N-1)}$$

$$s_{f1}^t = \left(\frac{N-2}{N-1}\right)s_f^t \tag{6.67}$$

The residual strength of the coupon in tension at the end of the tension part of the cycle is given by

$$\sigma_{al}^t = X_{f1}^t v_{f1}^t \tag{6.68}$$

Determine now what happens during the compression part of the first cycle.

First-Cycle Compression Portion

The starting configuration is the end configuration from the tensile portion of the cycle in Section 1.1. This means the following:

- The fibre volume is given by Equation 6.63.
- The strength of the fibres in compression is given by the strength at the last compressive cycle. If this is the first compressive cycle, the strength of the fibres is still X_f^c, the fibre static strength value.
- The standard deviation of the fibre strength is given by the standard deviation at the last compressive cycle. If this is the first compressive cycle, the standard deviation of the fibre strength is still s_f^c, the static value.

The applied stress on the fibres is given by a relation analogous to Equation 6.60.

$$(\sigma_{fa}^c)_1 = \frac{\sigma}{v_{f1}^t} \tag{6.69}$$

In a manner analogous to the tension part of the cycle, all fibres with strength less than $(\sigma_{fa}^c)_1$ fail. The fraction of fibres that failed f_1^c is given by:

$$f_{1new} = \mathrm{cdf}((\sigma_{fa}^c)_1, X_f^c, s_f^c) \tag{6.70}$$

The stress is redistributed to the remaining fibres that are now all loaded by the new stress:

$$(\sigma_{fc})_{new} = \frac{(\sigma_{fa}^c)_1}{1 - f_{1new}} \tag{6.71}$$

Repeat the process corresponding to Equations 6.69, 6.70 until convergence of f_{1new} to $f_{1c} < 1$ or all fibres fail.

Based on the converged value of f_{1c}, calculate new fibre volume:

$$v_{f1}^c = (1 - f_{1c})v_{f1}^t \tag{6.72}$$

Note that here the previous fibre volume is that at the end of the tension part of the cycle.

Calculate new fibre strengths. To do this, the cycles to failure N after the compression part of the cycle are computed from Equation 6.36 rewritten for the compression portion of the cycle:

$$N = -\frac{1}{\ln(1 - p_C)} \tag{6.73}$$

with p_C the probability that the applied stress σ is greater than the compressive strength of a specimen, given by

$$p_{C1} = \mathrm{cdf}\left(\underbrace{\frac{\sigma}{v_{f1}^c}, X_f^c, s_f^c}_{\substack{\text{quantities before} \\ \text{first cycle is applied}}} \right) \tag{6.74}$$

Then, if X_f^c is normally distributed, the residual strength follows a normal distribution with mean and standard deviation given by Equations 6.26a and 6.26b, respectively:

$$X_{f1}^c = \left(\frac{N-2}{N-1}\right) X_f^c + \frac{(\sigma_{fa}^c)_1}{(1-f_{1new}^c)(N-1)}$$

$$s_{f1}^c = \left(\frac{N-2}{N-1}\right) s_f^c \tag{6.75}$$

The residual strength of the coupon in compression at the end of the compression part of the cycle is given by

$$\sigma_{a1}^c = X_{f1}^c v_{f1}^c \tag{6.76}$$

Extension to Subsequent Cycles
At the end of cycle 1, determine

- X_{f1}^t, s_{f1}^t, respectively, the mean strength and the standard deviation of the fibre strength in tension;
- p_{T1} the fraction of specimens with strength lower than the applied tensile stress;
- N_1^t the cycles to failure under tensile load only (wear-out only);
- σ_{a1}^t the residual strength of the coupon in tension after the tension part of the cycle;
- X_{f1}^c, s_{f1}^c, respectively, the mean strength and the standard deviation of the fibre strength in compression;
- p_{C1} the fraction of specimens with strength lower than the applied compressive stress;
- N_1^c the cycles to failure under compressive load only (wear-out only);
- σ_{a1}^c the residual strength of the coupon in compression after the compression part of the cycle.

These become inputs for exactly the same procedure as for cycle 1 but now for cycle 2 and subsequent cycles. Failure is obtained if all fibres fail or if the residual strengths in tension or compression for the coupon, σ_{a1}^t, σ_{a1}^c, respectively, equal the applied tensile or compressive stress during the cycle. Note that with this procedure, p and N change during each cycle.

6.6.5.2   Comparison to Test Results

In order to validate the model, vacuum-infused unidirectional carbon/epoxy specimens were fabricated and tested under static and fatigue loading. The static tests were used to determine the mean and the standard deviation of the static strength distributions in tension and compression. Coupons 100 mm long × 15 mm wide, with two different thicknesses 2.5 and 1.5 mm, were cut from a larger panel, and fibreglass tabs with taper near the test section to minimise stress concentration effects were bonded on for load introduction. The resulting test section in each specimen was 10 mm long.

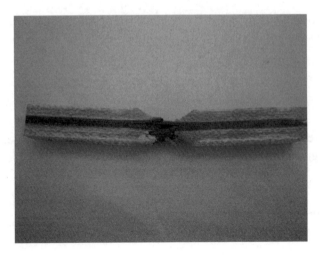

Figure 6.22   Unidirectional coupon used in static and fatigue tests (*See insert for colour representation of this figure.*)

Table 6.4   Static test results for unidirectional vacuum-infused specimens

	Tension	Compression
Mean (MPa)	1454	927.7
Standard deviation (MPa)	76.4	70.3

A typical specimen after static compression failure is shown in Figure 6.22. The average fibre volume was 49.5%.

Results from static strength tests are summarised in Table 6.4.

Analytical predictions, obtained using the approach in the previous section, are compared to the test results in Figure 6.23. There is good agreement between model predictions and test results. Further improvements can be made by incorporating effects of fibre waviness, variations in fibre diameter and fibre kinking. These would modify the strength predictions during each cycle and should give even better agreement with test results.

6.6.6   Application: Tension–Tension Fatigue of Cross-Ply Laminates

The case of $[0_m/90_q]$s cross-ply laminates is discussed qualitatively in this section. In the previous example of unidirectional laminates, the damage took mainly the form of fibre breakage, neglecting minor fibre splitting appearing sometimes towards the end of the fatigue life. In the case of cross-ply laminates, the damage is significantly more complex.

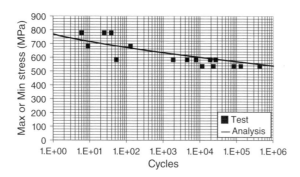

Figure 6.23 Comparison of analytical predictions to test results for tension–tension fatigue of unidirectional specimens

As mentioned in Sections 6.1 and 6.4, see Figure 6.11, damage first manifests itself as transverse matrix cracks in the 90° plies that extend to the interfaces with the 0° plies. As more load cycles are applied, these cracks multiply and the crack density increases. During this process, axial load in the 90° plies must be transferred around these cracks into 0° plies. This is done through interlaminar shear and normal stresses that develop at and near the 0/90 ply interfaces. As a result, the axial load in the 0° plies increases and the interlaminar stresses at the 0/90 ply interfaces may cause delaminations when the crack density in the 90° plies is high.

Given these damage mechanisms, the analysis model must be able to (i) predict the onset of matrix cracking, (ii) determine where and when subsequent matrix cracks will appear, (iii) calculate interlaminar stresses at the 0/90 interfaces, (iv) apply failure criteria for in-plane failure of the 0 plies or onset of delamination at the 0/90 ply interfaces and (v) predict/track delamination growth with cycles.

Details of the approach can be found in Section 3.4 and in Ref. [29]. Here, only a brief, qualitative, description will be given. The first matrix crack in the 90° plies appears when the axial stress there exceeds the *in situ* strength of the material. It is important to use the *in situ* strength because as the number q of 90° plies increases the *in situ* strength decreases significantly [31]. Once the first matrix crack is present, the three-dimensional state of stress in its vicinity is determined by using the calculus of variations to minimise the complementary energy in the laminate as described in Section 3.4 and in Ref. [29]. Qualitatively, the resulting stresses are shown in Figure 6.24.

Looking at Figure 6.24a, the σ_x and τ_{xz} stresses are zero at the crack location in the 90° plies. The interlaminar shear stress τ_{xz} increases from zero to a maximum value and then goes asymptotically back to zero. The σ_x stress increases from 0 to its far-field value that would be measured when there is no crack present. The interlaminar normal stress σ_z is compressive at the crack and decays to zero.

A slightly different picture is seen in Figure 6.24b for the 0° plies, in particular, with regard to the σ_x stress. In this case, σ_x starts from a high value, because the axial load

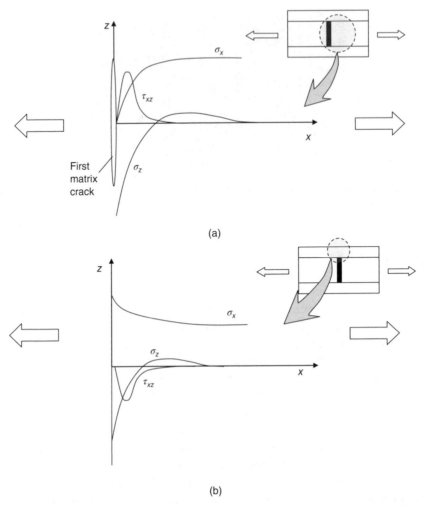

Figure 6.24   Stress distribution in the vicinity of one matrix crack in a $[0_m/90_q]$s laminate. (a) Stresses at mid-plane of 90° plies (only one crack present). (b) Stresses at mid-plane of 0° plies (only one crack present in 90° plies)

in the 90° plies has been transferred to the 0° plies right at the crack location. Then, as the axial distance from the crack increases, σ_x decreases asymptotically to its far-field value (when there is no crack present). It should be noted that Figure 6.24 is not to scale. For example, the σ_x stress in the 0° plies is, typically, 5–10 times greater than in the 90° plies.

    Looking at Figure 6.24, the compressive σ_z stress is unlikely to cause delamination. If anything it delays it. Also, it turns out, the magnitude of τ_{xz} is not high enough to cause delamination at least at the beginning of cycling. Then, the only stress that can lead to failure is σ_x. During cycling, the residual strength of both 0° and 90° plies is reduced. In the 0° plies, this takes the form of fibre breakage as the fibre residual strength decreases, see, for example, the first of Equations 6.67. In the 90 plies, this takes the form of matrix cracks. The stress σ_x in the 0 plies, even though it is higher at the matrix crack location than its far-field value, is not high enough to fail the ply. It only causes failure of the weakest fibres in a scenario analogous to that presented in the previous section.

It is the σ_x stress in the 90 plies that will cause the next failure. The second matrix crack appears when the stress in the 90° plies away from the matrix crack exceeds the updated *in situ* transverse strength of the plies. Updating is done through equations analogous to Equations 6.67 but referring to the transverse strength of the ply. This allows determination of the location of the next crack as the distance away from the first crack over which σ_x^{90} equals 99% of its updated far-field value. Note that σ_x is also updated because as fibres in the 0° ply fail, the corresponding fibre volume changes diverting a small amount of load to the 90° plies. Usually, the applied cyclic stress is high enough that more than one crack appear in the 90° plies during the very first cycle.

Once the second matrix crack appears in the 90° plies, the stress distributions of Figure 6.24 change to those of Figure 6.25. Now, σ_x and τ_{xz} are zero in the 90° plies at both crack surfaces. The peak values for the τ_{xz} and σ_z stresses increase. The σ_x stress in both plies goes towards its far-field value of the uncracked case but, if the crack spacing is small, there is not enough room before it has to 'turn again' to go to its peak value for the 0 plies and zero value for the 90 plies. 'Not enough room' in this case means that the proximity of one crack affects the stress field around the other.

The situation upon further cycling is completely analogous to that in Figure 6.25 with the crack spacing increasing as more cracks form. The peak values of the different stresses increase. This means that the maximum tensile σ_z stress halfway between the matrix cracks may at some point exceed the interlaminar tensile strength of the 0/90 interface and cause a delamination. This happens, typically, towards the end of the fatigue life. So, as the crack density increases, one must check the residual strength of the 0 fibres at locations aligned with sites where the cracks form in the 90° plies, the transverse strength of the 90° plies half-way between every two cracks and the interlaminar failure strength of the 0/90 interface again at the mid-point between any two cracks. Final failure occurs when all fibres in the 0° plies fail.

Checking whether the local stress state causes failure because the stresses exceed the residual strength of the components of interest requires the use of a failure criterion. For in-plane stresses, the Puck [26] or the Larc03 [27] criterion may be used. For delamination creation and growth, an approach analogous to that in Ref. [24] can be fruitful but the use of a Paris-type law may not be better than the residual strength calculations presented in this chapter. The interaction of in-plane failure such as matrix cracks with delaminations can be a major challenge. For design purposes, and until more efficient simulation methods are developed, simplified approaches may be necessary.

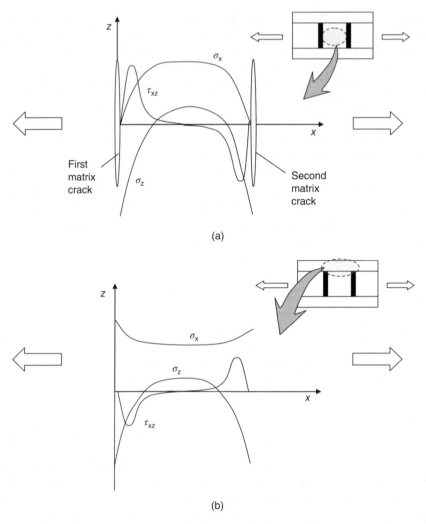

(a)

(b)

Figure 6.25   Stress distribution between matrix cracks in a $[0_m/90_q]s$ laminate. (a) Stresses at mid-plane of 90° plies (two or more cracks present). (b) Stresses at mid-plane of 0° plies (two or more cracks present)

Exercises

6.1   Looking at Figure 6.15, the high–low and low–high predictions are 'truncated' first when $n_1/N_1 = 1$ (the line drops vertically) and second when $n_1/N_1 =$ something less than 1 (0.6 is used in the figure as an example). Describe what happens physically in each case and why the two lines are 'truncated'.

6.2   Damage growth or crack growth in metals is characterised by a repeating pattern where the crack grows in one direction and, at any given time during the fatigue life, the crack can be 'scaled' from a previous time. This is self-similar damage growth. Discuss to what extent and under what conditions, if any, one can have self-similar growth in a composite. What types of damage or defect might exhibit self-similar growth?

6.3 Discuss how the damage scenario changes from that presented in Section 6.6 if, instead of $R = 0.1$, $R = -1$ is used. Pay particular attention to the creation of delaminations.

6.4 Assuming that the wear-out mechanisms in this chapter are the only fatigue mechanism in a composite structure, derive an 'exchange rule' that allows replacing a higher number of cycles at a lower load with a lower number of cycles at a higher load. This can be very useful during lengthy spectrum loading tests for reducing the test duration. (Hint: Use equations in Section 6.4.1.) In view of the discussion in Section 6.1, comment on the use of such an exchange rule.

References

[1] Hahn, H.T. and Kim, R.Y. (1975) Proof testing of composite materials. *J. Compos. Mater.*, **9**, 297–311.

[2] Vassilopoulos, A.P. and Keller, T. (2011) *Fatigue of Fiber Reinforced Composites*, Springer.

[3] Reifsnider, K.L. (ed) (1991) *Fatigue of Composite Materials*, Elsevier.

[4] Talreja, R. (1987) *Fatigue of Composite Materials*, Technomic Publishing.

[5] Yang, J.N., Lee, L.J. and Sheu, D.Y. (1992) Modulus reduction and fatigue damage of matrix dominated composite laminates. *Compos. Struct.*, **21**, 91–100.

[6] Lee, L.J., Yang, J.N. and Sheu, D.Y. (1993) Prediction of fatigue life for matrix-dominated composite laminates. *Compos. Sci. Technol.*, **46**, 21–28.

[7] Reifsnider, K.L., Sculte, K. and Duke, J.C. (1983) in *Long-Term Fatigue Behavior of Composite Materials* (ed T.K. O'Brien), ASTM, Philadelphia, PA, pp. 136–159, Long-Term Behavior of Composites, ASTM STP 813.

[8] Badaliance, R. and Dill, H.D. (1982) Compression Fatigue Life Prediction Methodology for Composite Structures. NADC-78203-60.

[9] Kassapoglou, C. (2011) Fatigue model for composites based on the cycle-by-cycle probability of failure: implications and applications. *J. Compos. Mater.*, **45**, 261–277.

[10] Mandell, J.F., Samborsky, D.D., Wang, L. and Wahl, N.K. (2003) New fatigue data for wind turbine blade materials. *Trans. ASME J Solar Energy Eng.*, **125**, 506–514.

[11] US Department of Defence and Federal Aviation Administration (1997) Composite Materials Handbook, Polymer Matrix Composites. Guidelines for Characterization of Structural Materials, vol. **1**, Chapter 8.3.4, Mil-Hdbk-17-1F, US Department of Defence and Federal Aviation Administration.

[12] Yang, J.N. and Jones, D.L. (1983) in *Load Sequence Effects on Graphite/Epoxy [±35]s Laminates, Long Term Behavior of Composites* (ed T.K. O'Brien), ASTM, Philadelphia, PA, pp. 246–262, ASTM STP 813.

[13] Wedel-Heinen, J., Tadich, J.K., Brokopf, C. *et al.* (2006) Optimat Blades- Reliable Optimal Use of Materials for Wind Turbine Rotor Blades. Report OB-TG6-R002, Chapter 13.2. https://www.wmc.eu/public_docs/10317_008.pdf.

[14] O'Brien, T.K. ASTM STP 775(1980) *Characterization of Delamination Onset and Growth in a Composite Laminate in Damage in Composite Materials*, American Society for Testing and Materials, pp. 140–167.

[15] Kassapoglou, C. (2007) Fatigue life prediction of composite structures under constant amplitude loading. *J. Compos. Mater.*, **41**, 2737–2754.

[16] US Department of Defense (2003) Metallic Materials and Elements for Aerospace Vehicle Structures, US Department of Defense, pp. 9–253, MIL-HDBK-5J, January 2003 Table 9.10.1.

[17] Kassapoglou, C. (2010) Fatigue of composite materials under spectrum loading. *Composites Part A*, **41**, 663–669.

[18] Broutman, L.J. and Sahu, S. (1972) A new theory to predict cumulative fatigue damage in fiberglass reinforced plastics, in *Composite Materials Testing and Design, (2nd Conference)*, American Society for Testing and Materials, pp. 170–188, ASTM STP 497.

[19] Gathercole, N., Reiter, H., Adam, T. and Harris, B. (1994) Life prediction for fatigue of T800/5245 carbon-fibre composites: I. Constant amplitude loading. *Fatigue*, **16**, 523–532.

[20] Gerharz, J.J., Rott, D., and Schuetz, D. (1979) Schwingfestigkeitsuntersuchungen an Fuegungen in Faserbauweise. BMVg-FBWT, pp. 79–23.

[21] Adam, T., Gathercole, N., Reiter, H. and Harris, B. (1994) Life prediction for fatigue of T800/5245 carbon-fibre composites: II variable amplitude loading. *Fatigue*, **16**, 533–547.

[22] Camanho, P.P., Dávila, C.G. and De Moura, M.F. (2003) Numerical simulation of mixed-mode progressive delamination in composite materials. *J. Compos. Mater.*, **37** (16), 1415–1438.

[23] Camanho, P.P., Dávila, C.G., Pinho, S.T. *et al.* (2006) Prediction of in situ strengths and matrix cracking in composites under transverse tension and in-plane shear. *Composites Part A*, **37** (2), 165–176.

[24] Turon, A., Costa, J., Camanho, P.P. and Dávila, C.G. (2007) Simulation of delamination in composites under high-cycle fatigue. *Composites Part A*, **38** (11), 2270–2282.

[25] Mohammadi, S. (2012) *XFEM Fracture Analysis of Composites*, Chapter 4.6, John Wiley & Sons, Inc., New York.

[26] Puck, A. and Schurmann, H. (2002) Failure analysis of FRP laminates by means of physically based phenomenological models. *Comps. Sci. Technol*, **62**, 1633–1662.

[27] Dávila, C.G., Camanho, P.P. and Rose, C.A. (2005) Failure criteria for FRP laminates. *J. Compos. Mater.*, **39**, 323–345.

[28] Qian, C. (2013) Multi-scale fatigue modelling of wind turbine rotor blade components. PhD Thesis. Delft University of Technology, Chapter 4.

[29] Kassapoglou, C. and Kaminski, M. (2011) Modeling damage and load redistribution in composites under tension-tension fatigue loading. *Composites Part A*, **42**, 1783–1792.

[30] De Backer, W. (2013) Development of an improved model for static analysis of unidirectional fiber reinforced polymer composites under compression. MSc Thesis. Delft University of Technology.

[31] Dvorak, G.J. and Laws, N. (1987) Analysis of progressive matrix cracking in composite laminates II. First ply failure. *J. Compos. Mater.*, **21**, 309–329.

7

Effect of Damage in Composite Structures: Summary and Useful Design Guidelines

Our understanding of the effects of damage on the performance of composite structures is far from complete. The complexity of the phenomena requires very detailed analytical models across many length scales supported by targeted tests that identify how damage is created and evolves and how larger scale properties such as stiffness and strength are affected. A lot of progress has been made in this direction but the numerical simulations that accompany the more advanced analysis models are not yet in a state where they can be applied to composite structures during design and preliminary analysis stages. To fill the need for analysis tools that can give reasonably accurate predictions and can be used in an optimisation environment, this book presented some candidate approaches for dealing with different types of damage in composite structures. As such, the methods have their limitations both in accuracy and range of applicability. Nevertheless, the material presented identified some basic trends and characteristics that can serve as useful design guidelines. They are summarised below. It is important to note that these guidelines are not absolute and may contradict each other in different situations. Still, they are very effective in narrowing down the field of options and selecting a robust design.

1. Holes
    (a) Including 45/−45 plies reduces the stress concentration factor caused by a hole. An all ±45 laminate has the lowest stress concentration factor that, for typical carbon/epoxy materials, is around 2.5. The strength, however, of an all ±45 laminate is low.
    (b) For optimum performance of a composite with a hole under uniaxial load, a certain percentage of plies with fibres aligned with the load must be present. For good balance of high strength obtained from the 0 plies (fibres aligned

Modeling the Effect of Damage in Composite Structures: Simplified Approaches, First Edition. Christos Kassapoglou.
© 2015 John Wiley & Sons, Ltd. Published 2015 by John Wiley & Sons, Ltd.

with the load) and low stress concentration factor, obtained from ±45 plies, the percentage of ±45 plies should be in the range of 20–40%. Plies of other orientations must also be used in accordance with the loads present and any other design guidelines such as the 10% rule.

(c) The stress concentration factor should not be used as a design tool as it is overly conservative. For the cases examined in Chapter 2, the strength of a laminate with a hole is at least 66% higher than what the stress concentration factor would predict. The stress concentration at the edge of a hole in a composite leads to the creation of a damage process zone with matrix cracks and delaminations, which limit the stress locally to values significantly below what the stress concentration factor would suggest.

(d) Finite-width effects are very important in composites with holes and they reduce the strength significantly below the strength of an infinite plate with a hole. A modified Whitney–Nuismer approach with analytically determined stress-averaging distance can give accurate failure predictions.

(e) It is recommended to use filled-hole tension and open-hole compression tests to establish tension and compression allowables in the presence of holes. It should be checked, however, whether the open-hole tension of the filled-hole compression tests are more conservative, which happens in a minority of cases.

2. Cracks
   (a) Through-thickness cracks in composites do not occur as frequently as in metals. Under typical repeated loads, they do not grow. Instead, a damage process zone develops ahead of the crack tip consisting of matrix cracks and delaminations, limiting the stress to a finite value. When they do grow, through-the-thickness cracks rarely grow in a self-similar manner.

   (b) The stress singularity at the crack tip of a composite is significantly lower than that in a metal. For typical carbon/epoxy materials, it is in the range of 0.2–0.3 as opposed to 0.5 in metals. Using linear elastic fracture mechanics in composites with a 1/2 stress singularity leads to unconservative predictions.

   (c) The finite-width correction for cracks is lower than for holes. From this point of view, cracks are less severe than holes. Holes are more conservative in establishing allowables for limit load capability of a composite structure.

   (d) Cracks may be analysed as holes with the diameter equal to the projected crack length perpendicular to the loading direction.

3. Delaminations
   (a) A composite structure with a delamination must meet ultimate load. It is important to relate the ultimate load capability in the presence of a delamination with the smallest delamination size that the inspection method selected can detect.

   (b) Inspection intervals must be established such that a delamination will not grow to a critical size during at least two inspection intervals. Critical size here refers to the ability to meet limit load if the delamination is detectable after two inspection intervals.

(c) Buckling of delaminating sub-laminates, wherever applicable, can be used as a conservative method to design against delamination growth for in-plane loading situations. As a rule, an embedded delamination under in-plane loads will not grow before buckling of at least one sub-laminate is reached.

(d) For general geometries under combined loading, delamination growth can be quantified using the strain energy release rate, or modelling with cohesive zone elements.

(e) Delamination is confined in the thin resin layer between plies. Growth takes place inside that layer which is homogeneous and isotropic. Thus, within the context of fracture mechanics, the strength of the singularity at the delamination edge is 1/2. This should be juxtaposed with the less than 1/2 singularity for through-the-thickness cracks, see summary point 2(b).

(f) Minimising the interlaminar stresses at free edges of composite laminates minimises the tendency to delaminate. This means minimising the differences in angle between plies (related to mismatch in Poisson's ratio and coefficient of mutual influence).

(g) In order to delay delamination onset and growth, materials with toughened resins, with high G_{Ic}, G_{IIC} or G_{IIIC}, should be used. This includes thermoplastic materials. Alternatively, stitching has the same effect in increasing G_{IC}.

4. Impact

(a) Composite structures with impact damage at or below the threshold of detectability must meet ultimate load. If visual inspection is used as the detection method, the corresponding impact damage is barely visible impact damage.

(b) Damage resistance, the ability to minimise damage for a given impact level, does not directly translate to damage tolerance, the ability to meet an applied load given a certain type and size of damage. While the two concepts are related, damage tolerance must be evaluated separately to ensure the required residual strength after impact.

(c) Modelling impact as a hole or a delamination of equivalent size can help predict compression or shear after impact for certain laminates, in particular sandwich with up to 12 ply facesheets and layup close to quasi-isotropic. For more general models, the effect of damage created during impact must be modelled accurately.

(d) Using tougher resins improves damage resistance because the delaminations created during impact are smaller. However, if the requirement is to meet ultimate load with barely visible impact damage, higher impact energies are required to cause barely visible damage when tougher resins are used. Higher energies lead to more fibre fractures at the impact site, which, in turn, may reduce the compression after impact strength.

(e) For improved compression after impact, the damage pattern created during impact must create relatively thick sub-laminates with little or no damage, in particular delaminations. These sub-laminates are the main load-carrying members during compression and have high enough bending stiffness to delay delamination buckling. This means that, in order to partly absorb the impact energy through delamination creation, it is acceptable to create relatively large delaminations away from these sub-laminates.

(f) Minimising the angular change from 1 ply to the next in a laminate reduces the size of the created delaminations at the respective interface and increases compression after impact strength.

(g) Softer laminates absorb more impact energy by bending. This helps minimise the damage created. However, softer laminates may have lower in-plane strength. An optimum combination of stiffness and strength must be selected for best compression after impact performance.

(h) Even though delaminations during impact start to form nearer the impact surface, for high enough impact energies, the biggest delaminations will be near the mid-plane. This means that the sub-laminates that should be created for improved compression after impact (CAI) as mentioned in point 4(b) should account for that and locate some 0 ply near the mid-plane along with clusters of angles with small differences from 0 so delamination creation is limited and buckling resistance of sub-laminates is increased.

(i) Modelling the impact damage region as consisting of inclusions with different stiffnesses representing the local damage present allows accurate predictions of the local stresses. These can be combined in a good progressive failure criterion to obtain accurate compression after impact predictions.

(j) Accurate progressive failure models and failure criteria with updated properties after damage are needed for predicting the extent of damage during impact and compression after impact strength.

(k) If the delaminations created during impact are sizeable, sub-laminate buckling may be the main mechanism precipitating compression after impact failure. Understanding the interaction between strength in the presence of damage and sub-laminate buckling is very important for obtaining accurate residual strength predictions.

5. Fatigue of composites

(a) Fatigue of composites occurs over many length scales. Understanding the creation of damage and its evolution requires the creation of models that can bridge these different scales.

(b) No matter which length scale is used in the analytical model as the 'starting' scale for modelling damage, there will always be lower length scales where damage will also occur and cannot be captured by the model at longer scales. A wear-out model that tracks macroscopic properties such as stiffness and/or strength will be needed. In addition, tests that capture lower scale damage creation and evolution will serve as the starting point of the model.

(c) Residual strength can be used to track how a composite structure evolves under fatigue loading. Models for residual strength based on the state of the structure at any given time relate the applied cyclic load, the number of cycles and the cycles to failure under the current load with the residual strength.

(d) Damage growth as a result of fatigue loading is rarely self-similar in composites. It consists of a combination of matrix cracks, delaminations and fibre breakages. These combine in various ways as a result of applied loading, geometry and stacking sequence used.

(e) The probability that the applied cyclic load exceeds the strength of a specimen can be related to the cycles to failure of a composite if there are no changes in the failure mechanism and provided that damage evolution occurs at low length scales not captured by the explicit analytical models used for each situation.

(f) Accurate fatigue analysis models must combine models that can predict the residual strength of a composite structure given its current damage state with models that can estimate the degradation of structural properties as a result of processes at lower length scales.

(g) Knowledge of the evolution of scatter of residual strength during fatigue testing can be very useful in determining the fatigue life.

Index

average stress, 21, 23, 151
averaging distance, 23–29, 222

barely visible impact damage, 107, 116, 196, 223
bearing, 9–12, 33
bearing by-pass, 11, 12
bearing stress, 11, 12, 33
block (see load block)
boundary conditions, 51, 64, 65, 70, 71, 102, 112, 126, 130, 138, 180, 182
   simply supported, 36, 64, 65, 69, 70, 72, 102, 126, 128
   clamped, 64, 70, 72, 73, 102, 123
bridging, 58, 59
BVID, 107, 163–165, 167, 196
bypass, 11, 12
   load, 12
   strain, 12

CAI, 167, 198, 224
calculus of variations, 129, 215
Castigliano's second theorem, 94
characteristic distance, 11, 23, 25, 28, 33, 37, 97, 122, 144
coefficient of mutual influence, 57, 75
coefficient of variation, 185
cohesive elements, 96, 99, 100
coin tap, 61, 100, 103

complementary energy, 77, 78, 89
complex elasticity, 13, 20, 24
compliance, 52, 88, 89, 127, 136
compression after impact, 109, 158, 198, 223, 224
concave tool, 58
contact pressure, 128, 129
contact radius, 128, 137, 164
contour mismatch, 59
crack, 41, 42, 156, 157, 215, 225
crack closure, 92, 93
crack growth, 41, 42, 75, 173, 218
   self-similar, 173, 219
crack (matrix), 49, 105, 115, 125, 215
critical size (delamination), 60–62, 222
critical energy release rate, 63, 79, 81, 87, 88, 92, 97
C-scan, 146–148, 163
cumulative distribution function, 185, 210

damage, 1, 97, 98, 105, 173, 225
   accumulation, 176, 180, 183
   manufacturing, 105, 106, 108, 173
   onset, 128, 138, 143, 175, 209
   resistance, 109
   scales, 2, 19, 41, 44, 175–177, 180, 221, 225
   service, 9, 60, 69, 105, 106, 108

Modeling the Effect of Damage in Composite Structures: Simplified Approaches, First Edition. Christos Kassapoglou.
© 2015 John Wiley & Sons, Ltd. Published 2015 by John Wiley & Sons, Ltd.